AIR CONDITIONING & REFRIGERATION TECHNICIAN's

EPA CERTIFICATION GUIDE

☑ *Getting Certified*
☑ *Understanding the Rules*
&
☑ *Preparing for EPA Inspections*

James F. Preston
Alias *"FREDDIE FREON"*

D-AMP PUBLICATIONS
Moon Township, PA

Air Conditioning & Refrigeration Technician's
EPA CERTIFICATION GUIDE

Getting Certified, Understanding the Rules,
& Preparing for EPA Inspections

By James F. Preston

Copyright © 1994 by James Preston

Printed and bound in the United States of America on acid free paper

Disclaimer of All Warranties and Liabilities

Publisher's - Cataloging In Publication
(Prepared by Quality Books Inc.)

Preston, James F.
    Air Conditioning & Refrigeration Technician's EPA Certification Guide: getting certified, understanding the rules, & preparing for EPA inspections / James F. Preston.
    p. cm.
    Includes bibliographical references and index.
    Preassigned LCCN: 94-071513.
    ISBN 0-943641-10-1

    1. Chlorofluorocarbons--Environmental aspects--Laws and Legislation--United States. 2. Refrigerants--Environmental aspects--Law and Legislation--United States. **I.** Title.

    TD887.C47P74 1994   363.7'38'088697
                         QBI94-993

For information on distribution or quantity discount rates, tel. voice/fax 412/262-5578 or write to: Sales Department, D-AMP PUBLICATIONS, 401 Amherst Avenue, Moon Township, PA 15108. Library distributors include Unique Books, Quality, and Baker & Taylor. Also available through wholesalers.

TABLE OF CONTENTS

FOREWORD

Thirty or forty years ago, somebody coming out of high school but not wanting to go to college would work in one of the "trades." They would be machinists, press operators, mechanics, carpenters or HVAC/R technicians. Machinists still make good money, but more and more machines are operated by computers, not skilled people. Carpenters are still in demand, but these days carpenters work with a powered nail gun, not a hammer. All of the "trades" are under some sort of pressure, be it technologically, changes in consumer demand, or increasing regulation.

Since mechanical refrigeration's invention, there hasn't been much change in the physical aspects. You still compress a gas and it absorbs heat as it expands. There could be said to be only two really big discoveries in refrigeration. The first is the invention of Freon and other artificial refrigerants. Artificial refrigerants meant that an appliance with an odorless, lighter than air gas could chill food in the kitchen, or be propped in the window to chill a room. After a while, every home and car had an air conditioner.

The second discovery is the thesis that artificial refrigerants pose a long term danger to the environment. Thus began the biggest challenge in refrigeration history: the reeducation of thousands of skilled persons who had counted on a life of honest work and honest money, without much interference. And believe me, some of them aren't happy about it.

Many technicians are going to avoid getting the required certification. Unfortunately, their employers can't keep them. The self-employed put themselves at risk by not complying. It may be the idea of big brother telling you what you can and cannot do. It may be the fear of failing the test. It may be procrastination. But this is not an option for the technician - you must take the certification exam if you want to keep your job!

Anybody that graduated from high school can probably pass the test. But, preparation is the key to anything, even if you've been in the field a long time. The *Air Conditioning & Refrigeration Technician's EPA Certification Guide* should be a strong background. Thorough study of the book and completing *all* of the sample questions will prepare many technicians for the test. The book and a competent review class will get almost anyone through it. Like filing your 1040s, this test is an unpleasant but vitally necessary task that you'll be happier when you complete. And like your 1040s, you don't want to know what happens if you miss. You'll do great if you work this book, work the sample tests and keep an open mind. Good Luck!

Joe Cronley, Publisher, REFRIGERATION News, Atlanta
Certified Technician, Classes I and II

CHAPTER ONE

THE RULES ARE CHANGING

Will you be out of work on November 14th, 1994?

Federal law now requires all non-automotive air conditioning and refrigeration service technicians to be certified by the Environmental Protection Agency (EPA) by November 14, 1994.[1] EPA-approved certifying organizations are administering exams nation-wide. If you don't have a *CERTIFICATION CARD*, you can't work with freon! Since August 13, 1992, automotive air conditioning technicians are required to have EPA certification under Section 609 of the Clean Air Act.

To purchase class I or class II refrigerants, *IDENTIFICATION CARDS* issued by EPA-approved certifying programs under Section 608 of the 1990 Clean Air Act regulations **WILL BE REQUIRED by NOVEMBER 14, 1994**. Just to ease your mind, Certification card and Identification card are one and the same.

---

[1] EPA Stratospheric Ozone Protection, FACT Sheet, May 1993, and 40 CFR Part 82 - Protection of Stratospheric Ozone, Subpart F - Recycling and Emission Reduction, Sec. 82.161.

Since November 15, 1992 class I or II motor vehicle air conditioner refrigerants, CFC-12 or R-12, may not be sold or distributed in containers smaller than twenty (20) pounds unless the purchaser is certified, reference 40CFR part 82.34. Automotive technicians **MUST NOW** be certified under part 609 requirements.

HELP IS HERE

This book will assist companies and technicians to comply with the 1990 Clean Air Act's refrigerant recycling rule for sections 608 and 609 and prepare service technicians for the certification test. Companies can use this guide to plan for on-site EPA inspections, develop training plans, prepare for certification testing, and review EPA and DOT labeling and record keeping requirements. It also provides information on standards and practices for freon recovery and reprocessing, and information on prohibitions that will help you and your company avoid heavy EPA fines.

When technicians require refrigeration and air conditioning theory review, use other references such as **The Refrigeration Service Engineers Society's** *Service Application Manual.* You will also find a copy of EPA inspector check lists, hot line and bulletin board contacts, listings of EPA regional offices, testing organizations, sample certification tests, compliance dates for recycling rules and other requirements of the law. Appendix II lists approved technician certification programs for sections 608 and 609.

Learn how to effectively prepare for on-site EPA inspections including the new record keeping requirements, and where to locate equipment manufacturers, refrigerant suppliers and freon substitutes. Reprints of the appropriate federal registers are included for your convenience and you are provided other important references from EPA pamphlets, circulars and reports.

WHAT THIS BOOK WILL DO FOR YOU

This book is written for two categories of readers, the company owners and their staff that must prepare for the new rules, and for those who must learn the requirements of, or pass a certification examination on, *Part 82-PROTECTION OF STRATOSPHERIC OZONE Sections 608 and 609 of the CLEAN AIR ACT of 1990.* EPA calls the program the *"STRATOSPHERIC OZONE PROTECTION PLAN".* Others may refer to it as the *FREON RECOVERY ACT* or *"PART 608"* or *"PART 609".*

Hundreds of sample test questions (with correct answers) are presented and each question is referenced to the appropriate sections of the Freon Recovery Act. Relevant sections of the new law are included in this guide for quick reference. The EPA is providing 500 questions to approved certification examining programs. We have included hundreds of sample questions in various formats to help you retain the information. Remember, passing the certification test is required for you to continue working with refrigerants. Your job is on the line if you can't service your customers' air conditioning and refrigeration equipment. It is also important to know and retain this information if and when an EPA inspector visits your shop and asks questions about recovery/recycling or requests a practical demonstration.

PENALTY ANYONE?

In 1993 undercover EPA agents purchased small cans of R-12 refrigerant from 12 businesses.[2] Those unlucky businesses were fined over $170,000.00 for not checking for required certification cards. EPA investigators will be inspecting businesses and technicians for some of the following:

- Proper certification.
- Documentation of purchases of type I and type II refrigerants.
- Quantities of refrigerants purchased and used.
- Proper labeling of containers.
- Whether recovery/recycling equipment is used and registered.
- Whether you or your technicians demonstrate proper procedures for recovering and/or recycling refrigerant.
- Whether service hoses have shutoff valves within 12 inches of the service ends.
- How empty containers are disposed of.

DID YOU REGISTER?

All recovery/reclaiming equipment must be registered with the EPA effective August 12, 1992.

ISN'T MY BUSINESS TOO SMALL TO BE INSPECTED?

You say I don't have to worry? I'm too small for inspection — a little fish in the big ocean — EPA will never find me! ***Don't bet your job or business on it***. Both large and small businesses will be inspected. EPA is and will be checking the registration records that they receive for recovery equipment, records of refrigerant sales, the *Yellow Pages*, local advertising media and local businesses. Local businesses? That's right, local businesses. Hotels, meat processors, commercial office buildings etc, will be checked because owners/operators are now required to keep records on air conditioning and refrigeration equipment that use 50 pounds or more of refrigerant. The servicing organization is required to document the amount of freon added. If servicing required evacuation of freon from the system or parts thereof, the EPA will check to see that recovery/recycling equipment was used and if the servicing organization possessed such equipment on that service date.

---

[2]Rules updated, March 1993.

Lots of ways to FIND YOU and lots of ways to FINE YOU!!!

There are lots of ways to **FIND YOU** and lots of ways to **FINE YOU** for non compliance. EPA is performing random inspections, responding to tips and they **WILL PURSUE** cases against violators. Can you or your business afford the maximum fine of *$25,000.00 a day per violation* of EPA regulations or the court costs to contest whatever fines that might be assessed against you or your business?

> ### THE BOUNTY PROVISION
>
> EPA can pay an award, not to exceed **$10,000**, to any person who furnishes information or services which lead to a criminal conviction or a civil penality assessed as a result of a violation of the CAA.

CERTIFICATION TESTING ON SECTION 608

All test questions are provided by the EPA to EPA-approved testing programs. For part 608, the EPA must provide a pool of 500 questions divided into four groups; a core group and three technical groups, reference 40CFR 82.161 (b). A minimum of 25 questions from the core group and a minimum of 25 questions from each relevant technical group are used for the test. An EPA FACT SHEET dated May 1993 for developing a 608 Technician Certification Program identifies the test bank of questions as divided into two groups, rather than four groups. Group I contains general questions on ozone depletion, general industry practices and the laws and regulations affecting technicians. Group II contains sector-specific questions about leak-detection, recovery techniques, shipping, labeling and disposal requirements.

There are a minimum of 25 questions from Group I and a minimum of 25 questions from each of the appropriate sector-specific categories in Group II. Group I is the CORE group and Group II contains the three TECHNICAL Groups.

GRANDFATHERING OF SECTION 608

Many technicians studied for certification in groups and others attended trade association sanctioned manufacturers seminars to become certified on recovery/reclaiming techniques. Much of this training was conducted before the requirements of the law were readily available and in some cases before EPA rules were implemented.

EPA plans to *GRANDFATHER* individuals who participated in training and testing programs **PROVIDED** the testing programs were: Approved by EPA and, provided additional EPA-approved materials or testing to these individuals to ensure that the required level of knowledge was obtained.

If you participated in a *NON-EPA Approved Certification Training Program* you should review Appendix 2 to see if that program is now approved. If the training program that you participated in is now an approved EPA program, contact the training program office for *GRANDFATHERING* procedures. If you were part of a large group not affiliated with any of the listed programs you may want to contact the sponsor of your group to see if they will apply for EPA approval by following the process in Part 82, Subpart F., 82.161 (g).

GRANDFATHERING UPDATE

EPA's Stratospheric Ozone Hotline announced on March 25, 1994 that technicians who participated in a voluntary certification program were sill being considered. If those programs aren't grandfathered by the 11/14/94 cutoff, technicians will be allowed an additional six to eight months to become certified by an approved program. Additional guidance is expected as the November 14th deadline approaches.

TYPES OF SECTION 608 CERTIFICATION

Technicians covered by section 608 are defined by EPA as TYPE I,II,III and UNIVERSAL technicians.

- Type I technicians:
  Maintain, service or repair small appliances.

- Type II technicians:
  Maintain, service, repair or dispose of high or very high pressure appliances, except small appliances and motor vehicle A/C systems.

- Type III technicians:
  Maintain, service, repair or dispose of low-pressure appliances.

- Universal technicians:
  Maintain service or repair low- and high-pressure equipment, including small appliances. Universal technicians may work on ANY equipment EXCEPT motor vehicle air conditioning systems.

Note: Section 608 allows certified technicians under section 609 to work on motor vehicle-like appliances but section 609 does not make allowances for type II or Universal technicians to work on motor vehicle air conditioners.

CERTIFICATION TESTING ON SECTION 609
(MOTOR VEHICLE AIR CONDITIONERS)

Effective August 13th, 1992, motor vehicle servicing technicians may not perform service on a motor vehicle air conditioner refrigerant system (1) without properly using equipment pursuant to 82.36; (2) Unless such person has been properly trained and certified by a technician certification program approved by the EPA Administrator pursuant to 82.40.

Section 609 does not specify technician types similar to section 608. One certification fits all for (MVACs) Motor Vehicle Air Conditioners. The test must cover the following:

- Recommended service procedures for the containment of refrigerant
- Service procedures for use of extraction/recycling equipment
- Standard for MVACs refrigerant purity
- Anticipated future technological developments
- HFC-134a replacement for R-12
- Environmental consequences of freon release
- Adverse effects of stratospheric ozone depletion
- SAE j standards
- Other requirements of section 609

GRANDFATHERING SECTION 609

The EPA has approved twelve technician certification programs as of Sept 14, 1993. Several programs existed before the EPA approval process. The EPA called them VOLUNTARY programs and three initially met EPA requirements and were approved. Programs with formal EPA approval as technician training and certification programs as well as their effective approval and GRANDFATHERING dates are listed in Appendix 2. The twelve approved programs open to the public as of September 14, 1993 are by the:

√ Mobile Air Conditioning Society
√ International Mobile Air Conditioning Association
√ National Institute of Automotive Service Excellence
  C.F.C. Reclamation And Recycling Service, Inc.
√ The Greater Cleveland Automobile Dealers' Association
  Mechanic's Education Association
  New York State Association of Service Stations and Repair Shops, Inc.
  Rancho Santiago College
  Refrigerant Certification Services
√ Snap-on Tools Corporation
√ Texas Engineering Extension Service
√ Waco Chemicals, Inc

√ DENOTES GRANDFATHERING

There are other programs approved but not listed. They are only open to employees of those companies. Approval means that if you participated in a training program with one of the twelve above mentioned organizations your certification may meet the requirements for certification GRANDFATHERING by the August 14, 1994 deadline. Check Appendix 2 to confirm if the organization you trained with has GRANDFATHERING dates. Contact them if you attended training during accepted grandfathered training dates so that you can update your certification.

Programs must provide training and testing by one or more of the following.[3]

- On-the-job training
- Self-study of instructional materials
- On-site training with instructors
- On-site training with videos
- Hands-on demonstrations

JANUARY 1, 1996

"THE END OF REFRIGERATION AS WE KNOW IT?"

With few exceptions, no chlorofluorocarbons (CFCs) will be produced in the United States after December 31, 1995.[4] The price of refrigerants is increasing due to scarcity of supply and environmental taxes. Both of these factors make your refrigerant on-hand stock more valuable as supplies continue to decrease. Recycling and recovery will increasingly be important to the service industry to hold down costs. The Montreal Protocol was signed by 110 nations and it included the same December 31, 1995 CFC manufacturing cutoff date. The types of refrigerant that EPA is phasing out are; CFC-11, CFC-12, CFC-13, CFC-113, CFC-114, CFC-115, R-500, R-502, R-503.

The EPA Action Guide, *COOLING & REFRIGERATING WITHOUT CFCs*, suggests that companies develop action plans for managing refrigerant resources and planning for the replacement or modification of existing equipment as the supply of restricted refrigerants decreases. Eventually, even with recovery and recycling, the supply will be completely depleted. Alternative refrigerants are under development and R-134a is the current replacement for R-12. However, R-134a is not as efficient as R-12 and R-12 systems require modification.

---

[3]Reference 40CFR Part 82, Subpart B, Sec. 82.40.

[4]EPA Stratospheric Ozone Protection, ACTION GUIDE, revised September 1993.

The EPA Fact Sheet, *SHORT LIST OF ALTERNATIVE REFRIGERANTS*, lists proposed acceptable refrigerant alternatives for new chillers and refrigeration equipment and for retrofitting older equipment. By November 15, 1995, all substitute refrigerants will also have to be recycled.[5]

Another EPA fact sheet, *Resources, AIR CONDITIONING AND REFRIGERANTS*, revised Sept, 1993, lists trade and professional associations, air conditioner manufacturers whose equipment do not use CFCs, commercial refrigeration manufacturers, chemical manufacturers known to produce alternative refrigerants and other EPA Fact Sheets. Fact Sheets and other information are available from EPA by calling the Stratospheric Ozone Hotline 800-296-1996.

STRATOSPHERIC OZONE HOTLINE - (800/296-1996)

Section 608 establishes the federal regulations for:

- **CERTIFICATION REQUIREMENTS** for technicians, recycling and recovery equipment and reclaimers.

- **THE SALE OF** Hydrochlorofluorocarbons (HCFC) and Chlorofluorocarbons (CFC),

- **SERVICE PRACTICES** that maximize recycling during the servicing and disposal of air conditioning and refrigeration equipment.

- **SELF DISCLOSURE** to EPA that technicians have acquired recycling or recovery equipment and compliance with the rule.

- **ESTABLISHES SAFE DISPOSAL REQUIREMENTS** to ensure removal of refrigerants from apparatus that enter the waste stream with the charge intact.

---

[5]EPA, Recycling Refrigerant From Motor Vehicle Air Conditioners. July 1992.

CHAPTER TWO

CFC — THE DESTROYER

Chlorofluorocarbons, CFCs, were great discoveries and since the 1920s industry has found varying uses for these chemicals. They were used for solvents, processed into foam insulations, solid gaskets, packaging materials and coffee cups. Since the mid 1970s, industries have shifted to other non-polluting chemicals or use processes that bind the CFCs so they do not escape into the atmosphere. CFCs are best known as the culprits responsible for the destruction of Earth's OZONE layer.

When supersonic air travel began, scientists started studying the exhaust components and their chemical reactions with ozone in the stratosphere. The stratosphere starts at six miles above the earth and continues up to a maximum of thirty miles above the earth. Stratospheric air sampling, using balloons and satellite imaging, revealed the hidden ozone devastation. Closer analysis detected unexpected concentrations of CFCs and chlorine.

A HOLE IN THE TOP AND BOTTOM OF THE WORLD

Scientists throughout the world began intensive studies to determine the threat level and what could be done to stop ozone depletion. Satellite imaging revealed holes in the stratospheric ozone layer at both polar caps. Arguments still rage on about whether CFCs are the biggest factor causing the hole or if it is the Suns radiation and cold temperatures that are causing the depletion. The United States and other governments have taken a position that CFCs must be reduced, agreements were signed and federal laws were passed to support that position. Whether you are a service technician or a company that uses refrigerants, we must all comply with the laws that apply to us or incur very harsh fines.

The latest NASA report for 1993 shows a continued decrease of ozone resulting in a larger hole. Most Americans think the depletion is only at the North and South poles. NOT TRUE. NASA has detected a decrease in ozone as far south as the 45th parallel. That equates to Montreal, Canada, St. Paul, Minnesota, the Wyoming/Montana state line and Portland, Oregon. Increased levels of ultraviolet radiation have been shown to cause severe damage to vegetation and increased skin cancer. Can we wait for ALL factors to be identified to START on a course of action? I don't think we can afford to.

OZONE DEPLETION
WHAT YOU NEED TO KNOW

- Ozone molecules are made of three oxygen atoms in a row.
- Safe ozone is in the stratosphere, 6 to 30 miles above earth.
- Bad ozone is in the troposphere and contributes to smog and global warming.
- Ozone is produced by lightning and by the suns rays breaking up oxygen molecules.
- CFCs in the stratosphere break down from the sun's intense ultraviolet rays releasing a chlorine atom. The chlorine atom is very reactive and attacks the ozone molecule eventually breaking loose one of the oxygen atoms. The chlorine and oxygen atoms combine to form a molecule of chlorine monoxide. The ozone molecule is transformed into an oxygen molecule, DI-atomic oxygen.
- When an atom of oxygen comes near the chlorine monoxide molecule it attracts the oxygen atom from the chlorine atom forming a molecule of oxygen, with the chlorine atom looking for another molecule of ozone to attack. The one atom of chlorine may repeat the process 100,000 times. Chlorine monoxide has increased faster than predicted by natural processes and increases as ozone decreases.
- Reactive hydrogen and reactive nitrogen also destroy ozone.
- Chlorine depleting chemicals have been assigned numbers to measure their OZONE DEPLETION POTENTIAL or ODP.
- Ozone is safe when it blocks ULTRAVIOLET B rays from the sun.
- As ozone is destroyed, ultraviolet rays increase.
- Increased ultraviolet rays, UV-B, cause skin cancers, cataracts, and crop damage.
- The Montreal Protocol, regulating CFCs, was signed in 1987.
- A fancy name for an element that has two atoms is DI-atomic (oxygen).
- A fancy name for an element with three atoms is TRI-atomic (ozone).
- CFCs get in the stratosphere from wind currents.
- CFCs in the troposphere do not break down as fast because the UV-B is much less intense.

THE REST OF THE STORY
CFC's — OZONE & CANCER

Orbiting above the earth, an astronaut can look down on our home and see the thin blue ribbon that rims our planet. That transparent blanket — our atmosphere — makes life possible. It provides the air we breathe and regulates our global temperature. And it contains a special ingredient called ozone that filters deadly solar radiation. Life as we know it is possible in part because of the protection afforded by the ozone layer. Gradually, it has become clear to scientists and to governments alike that human activities are threatening our ozone shield. Behind this environmental problem lies a tale of twin challenges: the scientific quest to understand our ozone shield and the debate among governments over how to best protect it. Here is the story.[1]

OUR OZONE SHIELD — OZONE AND HUMANKIND

For nearly a billion years, ozone molecules in the atmosphere have safeguarded life on this planet. But over the past half century, humans have placed the ozone layer in jeopardy. We have unwittingly polluted the air with chemicals that threaten to eat away the life-protecting shield surrounding our world.

Although ozone molecules play such a vital role in the atmosphere, they are exceedingly rare; in every million molecules of air, fewer than ten are ozone. Nitrogen and oxygen make up the vast proportion of the molecules in the air we breathe. In this way, ozone resembles a critical spice in a pot of soup. Using just a few grains of a particular herb, a chef can season the whole pot with a distinctive flavor.

Ozone molecules show different character traits depending on where they exist in the atmosphere. About 90 percent of the ozone resides in a layer between 10 and 40 kilometers (6 and 25 miles) above the Earth's surface in a region of the atmosphere called the stratosphere. Ozone there plays a beneficial role by absorbing dangerous ultraviolet radiation from the sun. This is the ozone threatened by some of the chemical pollutants that we have released into the atmosphere.

Close to the planet's surface, however, ozone displays a destructive side. Because it reacts strongly with other molecules, it can severely damage the living tissue of plants and animals. Low-lying ozone is a key component of the smog that hangs over many major cities across the world, and governments are attempting to decrease its levels. Ozone in the region below the stratosphere — called the troposphere — can also contribute to greenhouse warming.

Although smog ozone and stratospheric ozone are the same molecule, they represent separate environmental issues, controlled by different forces in the atmosphere. This monograph will focus on the stratospheric ozone layer and the world's attempts to protect it.

---

[1]This chapter incorporates in part the monograph titled "Our Ozone Shield," Albritton, D.L., 1992, Reports to the Nation on our Changing Planet, No.2., with permission of the University Corporation for Atmospheric Research, Office for Interdisciplinary Earth Studies, Boulder, Colorado and the National Oceanic and Atmospheric Administration, Office of Global Programs, Silver Spring, Maryland.

What is ozone and where does it originate? The term itself comes from the Greek word meaning "smell," a reference to ozone's distinctively pungent odor. Each molecule contains three oxygen atoms bonded together in the shape of a wide triangle. In the stratosphere, new ozone molecules are constantly created in chemical reactions fueled by power from the sun.

The recipe for making ozone starts off with oxygen molecules (O_2). When struck by the sun's rays, the molecules split apart into single oxygen atoms (O), which are exceedingly reactive. Within a fraction of a second, the atoms bond with nearby oxygen molecules to form triatomic molecules of ozone (O_g).

Even as the sun's energy produces new ozone, these gas molecules are continuously destroyed by natural compounds containing nitrogen, hydrogen, and chlorine. Such chemicals were all present in the stratosphere — in small amounts — long before humans began polluting the air. Nitrogen comes from soils and the oceans, hydrogen comes mainly from atmospheric water vapor, and chlorine comes from the oceans.

The stratospheric concentration of ozone therefore represents a balance, established over the aeons, between creative and destructive forces. The total level of ozone in the stratosphere remains fairly constant, an arrangement resembling a tank with open drains. As long as the amount of water pouring in equals the amount flowing out the drain holes, the water level in the tank stays the same. In the stratosphere, the concentration of ozone does vary slightly, reflecting small shifts in the balance between creation and destruction. These fluctuations result from many natural processes such as the seasonal cycle, volcanic eruptions, and changes in the sun's intensity.

For about a billion years, the natural ozone system worked smoothly, but now human beings have upset the delicate balance. By polluting the atmosphere with additional chlorine-containing chemicals, we have enhanced the forces that destroy ozone — a situation that leads to lower ozone concentrations in the stratosphere. The addition of these chemicals is the same as drilling a larger "chlorine" drain in the tank, causing the level to drop.

A PROBLEM ARISES: THE EARLY 1970s

No one dreamed human activity would threaten the ozone layer until the early to mid-1970s, when scientists discovered two potential problems: ultrafast passenger planes and spray cans.

The plane threat surfaced first, after the invention of a new breed of commercial aircraft called supersonic transport (SST). These planes could fly faster than the speed of sound and promised to trim hours off long journeys. In the 1970s, the United States and other nations began considering whether to build large fleets of such ultrafast jets.

As scientists such as Harold Johnston and Paul Crutzen looked at the SST issue, they grew concerned about the effects such planes might have on the stratosphere. SSTs are unusual because they must fly high up in the atmosphere — where the air is thin — to achieve their fast speeds. Several researchers suspected that the reactive nitrogen compounds from SST exhaust might accelerate the natural chemical destruction of ozone, causing ozone levels to drop.

In 1974, news of another possible threat to the ozone layer made national headlines. This time scientists implicated a widely used class of chemicals known as chlorofluorocarbons

(CFCs), which were most commonly known as the aerosol propellant in spray cans. Invented in the late 1920s, CFCs contain chlorine, fluorine, and carbon atoms arranged in an extremely stable structure.

Through decades of use, CFCs proved themselves to be ideal compounds for many purposes. They are nontoxic, noncorrosive, nonflammable, and unreactive with most other substances. Because of their special properties, they make excellent coolants for refrigerators and air conditioners. CFCs also trap heat well, so manufacturers put them into foam products such as cups and insulation for houses.

Most scientists had not worried about how CFCs would affect the atmosphere. But two chemists, F. Sherwood Rowland and Mario Molina, began considering these wonder compounds, and they uncovered something disturbing. Because CFCs were extremely stable in the lower atmosphere, they could drift up into the stratosphere, where they would break apart when bombarded by the sun's high-energy radiation. CFCs therefore carried millions of tons of extra chlorine atoms into the stratosphere, adding much more than the amount of chlorine supplied naturally by the oceans in the form of methyl chloride.

Rowland and Molina hypothesized that the chlorine buildup from CFCs would spell severe trouble for the ozone layer. According to their predictions, each chlorine atom could destroy 100,000 ozone molecules, meaning that decades of CFC use could cause substantial declines in the concentration of stratospheric ozone.

Any drop in ozone levels, whether from SSTs or CFCs, would allow more ultraviolet light to reach the Earth's surface — an effect that holds severe consequences for life on the planet. Exposure to ultraviolet light enhances an individual's risk for skin cancer and cataracts, so an increase in this radiation could lead to more cases of such diseases. Ultraviolet light also harms food crops and other plants, as well as many species of animals.

Thus the world faced two ozone-related environmental issues in the first half of the 1970s. In terms of SSTs, policy makers had to decide whether to build such planes. With CFCs, the question was whether to limit the production and use of these chemicals.

Of all the countries considering SSTs, the United States had planned the largest fleet, and it addressed this issue rather quickly. When preliminary scientific studies suggested the planes would significantly thin the ozone layer, the U.S. government decided against the proposed fleet.

Political leaders faced a much tougher decision on the subject of CFCs. For example, in the United States, these extremely reliable chemicals formed the center of a multi-billion-dollar industry. Though the Rowland/Molina hypothesis warned that CFCs might endanger the health of the planet's inhabitants, officials feared that a ban on such chemicals would disrupt many segments of society. Was it worthwhile to face economic hardships solely because of a scientific hypothesis and its predicted effects?

Decision makers also knew that the ozone layer belonged to the entire world, meaning that all countries would have to address the problem.

STRATOSPHERIC OZONE: THE FIRST DECADE (1974 — 1984)

Would CFCs really bring significant harm to the ozone layer? That was the question politicians were asking in 1974, and the scientific community set out to provide an answer.

Atmospheric researchers had to judge the seriousness of the problem. If ozone levels were to decline by only 1 percent in the next 50 years, nations would have little cause for concern. On the other hand, a substantial drop in ozone levels could jeopardize the world.

The first attempts to assess the problem produced dire forecasts, suggesting that CFCs could destroy perhaps half the ozone shield by the middle of the next century. Yet experts did not know how much to believe these early estimates, because they were based on a very simplistic understanding of chemical reactions in the stratosphere.

It was like trying to decipher a partially completed jigsaw puzzle, spread out on a table. Scientists wondered what the missing pieces looked like and whether they would change the emerging picture.

Over the next few years, researchers took many different routes toward filling in the gaps in the ozone puzzle. Experiments in the laboratory allowed chemists to gauge how quickly chlorine destroyed ozone molecules. Other scientists launched balloons that carried instruments up into the stratosphere, where they measured the concentrations of key chemicals that controlled ozone levels. All this information fed into new computer models that predicted how chemicals would affect the ozone layer.

By 1976, many experts had grown convinced that CFCs did indeed present a serious threat. In the United States — the world's largest producer and user of CFCs — the public called for the government to place limitations on these chemicals. Civic leaders launched boycotts against items that used CFCs, and some companies even eliminated the compounds from their products.

The U.S. and some other governments responded in 1979 by banning the sale of aerosol cans containing CFCs. Because spray cans represented the largest use of these chemicals, the ban led to an abrupt leveling off of CFC production.

After the spray can decision, the ozone issue quickly receded from worldwide headlines. But atmospheric researchers knew that danger still threatened the protective ozone layer. While CFCs no longer filled U.S. aerosol cans, companies continued to produce these chemicals for use in air conditioners, in insulation, and in the cleaning of electronic parts. What's more, most countries aside from the United States continued to use CFCs in spray cans. So even as the threat to the ozone layer slipped from the public spotlight, scientists extended their investigations into the problem.

Researchers also began watching the ozone layer more closely, searching for evidence that chlorine pollution had already started weakening the protective shield. They knew it might be difficult to spot such destruction at first. Ozone levels fluctuate naturally by several percent, so identifying the subtle signs of unnatural ozone loss would be like trying to hear someone whisper a message across a crowded room.

The U.S. ban on CFC propellants in spray cans caused a temporary pause in the growing demand for the offending compounds. But worldwide use of the chemicals continued, and levels of CFC production began to rise again. By 1985, the production rate was growing 3 percent a year.

The increase in CFC use rekindled worldwide attention to the threat of ozone destruction, spurring countries in 1985 to sign an international agreement called the Vienna Convention. The convention called on negotiators to draw up a plan for a worldwide action on this issue. It also required scientists to summarize the latest information on the atmospheric consequences of CFCs and related bromine-containing chemicals called Halons, which had grown popular over the previous decade because of their ability to extinguish fires. Collectively, CFCs and Halons fit under the name halocarbons.

Using the most complete models, experts predicted that if levels of halocarbon production continued to increase as they had in the past, ozone concentrations in the stratosphere would drop by about 5 percent by the year 2050. Although much less severe than the predictions of earlier years, even a 5 percent decrease would still allow a very serious surge in the amount of ultraviolet radiation reaching the Earth's surface, causing millions of new cases of skin cancer in the United States alone.

By the time of the Vienna Convention, scientists remained uncertain whether ozone levels had actually started to drop. The research community, nonetheless, warned that countries could not afford to take a wait-and-see approach. Halocarbons present an insidious danger for the future because they can survive in the atmosphere for decades; some can last several centuries. That means even if the entire world stopped producing such compounds instantly, the halocarbons already in the atmosphere would continue to damage the ozone layer for more than 100 years. Many governments thought it critically important to limit the chemicals as soon as possible.

Then in May of 1985, shocking news spread throughout the scientific community. British researchers reported finding dramatic declines in ozone values over Antarctica each spring — actual "holes" in the ozone layer. Atmospheric scientists didn't know how to explain these large and unanticipated changes. Some proposed that natural processes were at work, while others thought it was the first sign that halocarbons were wearing away the protective ozone shield.

Despite uncertainty about the Antarctic phenomenon's cause, scientists firmly believed halocarbons would eventually deplete the global ozone shield. Their certainty and the jarring unexpectedness of the ozone hole's appearance motivated countries to act. In September 1987, diplomats from around the world met in Montreal and forged a treaty unprecedented in the history of international negotiations. Environmental ministers from 24 nations, representing most of the industrialized world, agreed to set sharp limits on the use of CFCs and Halons. According to the treaty, by mid-1989 countries would freeze their production and use of halocarbons at 1986 levels. Then over the next ten years, they would cut CFC production and use in half.

For scientists and policy makers, the Montreal Protocol marked a truly profound moment. When negotiators drew up the treaty, they were motivated by concerns about *future* ozone loss, rather than by direct observations of current ozone destruction by CFCs. (Certainly the ozone hole in Antarctica had unnerved world leaders, but it was by no means clear whether chemical pollutants had caused this decline.) Thus, the agreement was based primarily on confidence in a theory.

The Montreal Protocol established a new way of viewing environmental problems. In the past, the world had addressed such issues only after damage grew noticeable. For example, nations agreed to limit above-ground nuclear tests once it became evident these explosions poisoned the air and water with radioactivity. The Montreal agreement, however, tackled the ozone issue early, demonstrating a heightened sense of environmental responsibility.

The framers of the protocol also broke new ground in another way: they realized their agreement might not suffice if future scientific work revealed that the ozone layer faced even greater danger. Uppermost in their minds was concern over the Antarctic ozone hole and its possible implications for global ozone. The diplomats therefore included a provision calling for negotiators to reconvene in 1990 to examine any new scientific or technical information that might necessitate adopting deeper cuts.

THE OZONE YEARS: 1985 — 1989

The ozone hole was borne in the late 1970s, long before the Montreal Protocol was signed. Like a leak in the roof over the distant part of a house, the hole at first grew unnoticed by any human being living below.

Each spring, ozone abundances over the ice-covered continent dropped below normal and then rose gradually toward normal amounts in summer. And each year, the springtime losses grew worse.

A British team, which had measured ozone levels over the Antarctic coast since 1956, first began noticing the phenomenon in the early 1980s. But it was hard to swallow the evidence at first. Was the ozone hole real, or were the instruments malfunctioning? wondered the scientists. After checking and rechecking the instruments, the British researchers grew confident of their discovery. In 1985, they announced their startling news to the rest of the world.

Atmospheric experts moved quickly to determine whether the ozone hole was real. Consulting measurements made by satellite-borne and balloon-borne instruments, they found evidence confirming the springtime ozone depletion. Even more staggering, measurements showed the hole extending over the entire Antarctic continent.

The discovery of the ozone depletion blindsided the scientific community, catching it totally off guard and without a suitable explanation. But within a few months, theoretical scientists came up with three competing ideas that could explain why the ozone hole had developed over Antarctica.

One group of scientists focused on the solar cycle — the periodic waxing and waning of the sun's energy output. Noting that solar radiation had grown particularly strong in the early 1980s, some researchers proposed the intense radiation had created above-normal levels of reactive nitrogen chemicals in the stratosphere. These compounds could then concentrate over Antarctica and destroy ozone there.

A second group suggested that natural changes in stratospheric winds were responsible. According to this "dynamical" theory, the ozone hole resulted from changes in the system of air motions that transport ozone and establish its amount in the polar regions.

Both the solar cycle and dynamical theories stressed natural processes as a cause for the depletion. But a third theory held that human-made chemicals deserved the blame. According to this idea, the cold conditions above Antarctica amplified the ozone-destroying powers of CFCs and Halons, accelerating the loss in this region.

The three separate theories held profoundly different implications for the world. If halocarbon pollution created the hole, then scientists had gravely underestimated the chemicals' destructive power, and the ozone layer faced even more danger than previously thought. But if the hole formed because of natural processes, then humans could breathe a sigh of relief.

With very little known about the Antarctic ozone losses, atmospheric researchers could not tell which theory was correct. Yet they recognized that political leaders would need an answer as soon as possible. The signers of the Montreal Protocol would be meeting to review the limitations on halocarbons, and it was critical to know whether these chemicals lurked behind the ozone hole.

The scientific community threw itself at the problem, launching several field expeditions aimed at solving the riddle of the ozone depletion. In September of 1986, a hastily assembled team hurried off to McMurdo Station in the Antarctic. Using ground-based instruments and balloons to probe the stratosphere, this team found high levels of ozone-destroying compounds. A year later, the United States, in conjunction with other countries, sent a massive group of more than 100 scientists, engineers, and technicians to Punta Arenas, Chile, at the southern tip of South America. From this distant base, two research airplanes flew into the dangerously cold Antarctic sky to gather conclusive data about the mysterious affairs in the stratosphere over that icy land. Other scientists returned to McMurdo for further measurements.

By October 1987, the researchers came back from the Southern Hemisphere with a dark message for the world: blame for the ozone hole falls on human shoulders. The expeditions showed that chlorine and bromine pollution had shifted the fragile chemical balance in the Antarctic, thereby draining those skies of ozone during the spring.

Ozone loss is accelerated over the frozen continent because the Antarctic stratosphere contains cloud particles not normally present in warmer climes. These icy particles have a critical effect on the chlorine and bromine pollution floating in the stratosphere. Normally, the chlorine and bromine are largely locked into "safe" compounds that cannot harm ozone, but the ice particles transform them into destructive chemicals that can break apart ozone molecules with amazing efficiency. In 1987, ozone concentrations above Antarctica fell to half their normal levels, and the hole spread across an area the size of the United States.

Evidence gathered during these expeditions and new data from laboratories back home enabled scientists to fashion a consistent theory to explain the hole. In the prelude to ozone depletion, ice particles form during the polar night, when several months of darkness descend on Antarctica and temperatures plummet below 80°C (-112°F) in the stratosphere. On those floating ice particles, reactions convert chlorine from the "safe" to the "destructive" form. The real action begins when the sun returns to this part of the world during springtime, energizing the chemical cycle that destroys ozone. Wind patterns during winter and spring contribute by isolating the Antarctic stratosphere from warmer air to the north.

The ozone hole forms only in Antarctica because this region has a unique combination of weather conditions: it is the coldest and most isolated spot on Earth. But somewhat similar conditions exist in the Arctic, and scientists wondered whether the North also suffered from ozone loss. Even small depletions in this region would represent cause for concern, because many people live in northern latitudes potentially affected by Arctic ozone loss. So in 1988, two small teams traveled to Greenland and Canada to gather data. A year later, an extensive group headed to Norway to take measurements with the two airplanes that helped to solve the Antarctic puzzle.

The northern expeditions revealed that during wintertime, the Arctic stratosphere has the same types of destructive chlorine and bromine compounds that cause the problems in the Antarctic. Indeed, when scientists returned to the Arctic for an extended study in 1991 and 1992, they discovered strong hints that such compounds had destroyed significant amounts of ozone in the polar region. But because the Arctic atmosphere is not as isolated, the ozone losses there appear to be much smaller than those in Antarctica — at least for the present.

Between trips to the ends of the Earth, atmospheric scientists during this period also stepped up their search for signs of a global erosion in the ozone layer. An international panel of experts came together to scrutinize measurements made by satellites and by ground-based instruments around the world. In 1988, they reached a verdict: global ozone levels *had* declined over the past 17 years, mainly in the winter. Normal processes such as the solar cycle had caused part of the drop, but natural effects could not explain the entire ozone loss.

The news grew even worse. An international panel announced that ozone levels had dropped by measurable amounts not only in winter and spring but also in summer. Because people spend far more time outdoors during summer, ozone loss at this time of the year poses the greatest threat to the health of humans.

Scientists suspect that CFCs and Halons are to blame for much of the ozone decline, which has reached several percent over the mid-latitudes of the Northern Hemisphere — the segment of the globe that encompasses the United States and Europe. But atmospheric researchers are not yet fully confident that they know what mechanism lies behind the drop. The largest changes have occurred over the poles and neighboring midlatitudes, leading some researchers to suggest that loss near the poles has enhanced the decline in global ozone levels. Others suspect that the natural, thin layer of sulfur-containing particles in the stratosphere could be involved in midlatitude ozone loss, in a role somewhat similar to that played by ice particles over Antarctica.

The fast-paced research of the late 1980s revealed that the original Montreal Protocol would not go far enough toward protecting the fragile ozone layer. Even with the 50 percent cuts mandated by the treaty, levels of chlorine and bromine would still rise in the stratosphere, meaning that ozone loss would only worsen with time.

In June 1990, diplomats met in London and voted to significantly strengthen the Montreal Protocol. The treaty calls for a complete phaseout of CFCs by the year 2000, a phaseout of Halons (except for essential uses) by 2000, and a rapid phaseout of other ozone-destroying chlorine compounds (carbon tetrachloride by 2000 and methyl chloroform by 2005).

The treaty also attempts to make the phaseouts fair for developing countries, which cannot easily afford the higher-priced substitutes that will replace banned compounds. The revised agreement establishes an environmental fund — paid for by developed nations — to help developing nations switch over to more "ozone-friendly" chemicals.

OUR OZONE LAYER: PRESENT AND FUTURE

But many pieces of the ozone puzzle remain missing, and scientists wonder whether new ozone problems will develop in the near future. Experts are exploring several unanswered questions, including:

- What surprises lurk in the next decade or so? Even with the amended protocol, chlorine abundances will continue to rise until around the turn of the century.

- Will ozone losses grow worse in the Arctic as chlorine abundances increase?

- How safe are the CFC substitutes? Will some of them significantly contribute to ozone loss, global warming, or other environmental problems?

- How appropriate is it to allow countries to continue "essential" uses of the powerful ozone-depleting Halons? The current treaties permit these uses.

- Are there other compounds that significantly deplete the ozone layer and hence could deserve attention under the Montreal Protocol — such as methyl bromide, which is used widely as a fumigant?

- How will polar ozone destruction affect populated countries? Will the Antarctic hole cause ozone declines over Chile, Argentina, and New Zealand? Will Arctic losses spur drops in ozone concentration over Canada, Scandinavia, the United States and the former Soviet Union?

- How much do the natural particles in the stratosphere, other than the icy polar clouds, accelerate the chemical destruction of ozone at midlatitudes?

- How will large volcanic eruptions — which can inject immense amounts of dust into the stratosphere — affect the ozone layer when the chlorine from CFCs reaches unprecedented abundances?

- How will the ozone hole and global ozone losses affect worldwide weather and climate?

- Does a proposed new class of high-altitude aircraft threaten the ozone layer?

Decision makers will need answers to such questions as they continue to revisit their international agreements in the future and ask if these are adequate in light of new research findings.

The Montreal Protocol provides a dramatic example of science in the service of humankind. By quickly piecing together the ozone puzzle, atmospheric researchers revealed the true danger of halocarbons, allowing world leaders to take decisive action to protect the ozone layer.

This international agreement represents a critical step toward saving the world's ozone layer. But perhaps more importantly, it has taught scientists and policy makers an invaluable lesson about addressing environmental problems. Negotiations on this issue mark the first time the nations of the world have joined forces to protect the Earth for future generations.

The treaty can serve as a crucial apprenticeship for world leaders and scientists, who now face an even more daunting environmental matter — the threat of global greenhouse warming that looms over the future of this planet. The successful ozone agreement offers hope that scientific understanding can once again provide the foundation for responsible action by the international community.

SOME INTERESTING FACTS

✎ *Ozone is produced by solar radiation and lightning.*

✎ *Reactive chlorine destroys ozone.*

✎ *Reactive hydrogen destroys ozone.*

✎ *Reactive nitrogen destroys ozone.*

✎ *Ozone loss is occurring faster than production.*

✎ *Good ozone is in the stratosphere, blocking ultraviolet rays.*

✎ *Bad ozone is in the troposphere, causing smog and greenhouse warming.*

✎ *The oxygen molecule looks like two balls connected by a stick.*

✎ *Solar radiation breaks the oxygen molecule. The individual oxygen atoms are very reactive and quickly bond to each other, forming ozone. They could also bond to chlorine, hydrogen or nitrogen atoms and NOT form ozone. A three percent loss was detected from about Philadelphia to New Orleans.*

✎ *A four percent loss was detected from about the Canadian American border to Provo Utah and Philadelphia PA.*

✎ *Ozone is three atoms of oxygen. Imagine three balls connected by two sticks. Solar radiation breaks up CFC molecules producing reactive chlorine that bonds to an oxygen atom of the ozone molecule, causing the oxygen atom to release from the ozone molecule, destroying ozone.*

✎ *One chlorine atom may destroy up to 100,000 ozone molecules.*

This chapter incorporates in part the monograph titled "Our Ozone Shield", Albritton, D.L., 1992, Reports to the Nation on our Changing Planet, No.2., Fall 1992, with permission of the University Corporation for Atmospheric Office for Interdisciplinary Earth Studies, Boulder, Colorado and the National Oceanic and Atmospheric Administration, Office of Global Programs, Silver Spring, Maryland. For additional copies of the original monograph, contact the UCAR Office for Interdisciplinary Earth Studies, PO Box 3000, Boulder, CO, 80307-3000, phone: (303) 497-2692, fax: (303) 497-2699, internet: oies@ncar.ncar.ucar.edu.

CHAPTER THREE

PROHIBITIONS, RULES, REGULATIONS, & DATES

The Group I section of the certification test is a LOCKOUT item that concentrates on prohibitions, rules, regulations, and practices. If you fail the group I section of the test you fail the complete exam. There are twenty-five Group I questions derived from the EPA test pool of 75-100 questions. The Group I section includes questions on the environmental impact of CFCs and HCFCs, laws and regulations, and the changing industry outlook. The following information was taken from Section 608, 609 and 611 of the Clean Air Act. To prepare for the Group I test review this chapter's prohibitions, rules, regulations, and dates. Where Section 608 refers to Subpart B, Subpart B is Section 609, MVAC. Note that some dates have before and after requirements or specifications.

Key dates are provided with a short narrative for you to study in preparation for the Group I test.

SECTION 608, PROHIBITIONS PART 82.154, EFFECTIVE DATES

NOTE: Section 608 of 40 CFR 82.150 (b) applies to ANY PERSON who services, maintains or repairs appliances, except MVACs, or disposes of appliances INCLUDING MVACs, reclaimers, owners of appliances, manufacturers of appliances and recovery and recycling equipment. Also refer to 40 CFR 82.152 (x).

(a)   **June, 14 1993** — No person maintaining, servicing, repairing or disposing of appliances may knowingly vent or otherwise release into the environment any class I or II substance used as refrigerant in such equipment.

DE MINIMIS releases associated with good faith attempts to recycle or recover refrigerants are not subject to this prohibition. RELEASES SHALL BE CONSIDERED DE MINIMIS IF THEY OCCUR WHEN:

    (1)   The required practices set forth in 82.156 are observed and recovery/recycling machines meeting 82.158 are used OR

    (2)   The requirements set forth in CFR part 82, Subpart B are observed. (Section 609, part 82.34 and appendix A)

The knowing release of refrigerant subsequent to its recovery from an appliance shall be considered a violation of this prohibition.

(d)   **June 14, 1993** — No person shall alter the design of certified refrigerant recycling or recovery equipment in a way that would affect the equipment's ability to meet the certification standards set forth in 82.158 without   resubmitting the altered design for certification testing. Until it is tested and shown to meet the certification standards in 82.158, equipment so altered will be considered uncertified for the purposes of 82.158.

(b)   **July 13, 1993** — No person may open appliances except MVACs for maintenance, service or repair and no person may dispose of appliances except for small appliances, MVACs, and MVAC-like appliances: (1) Without observing the required practices in 82.156; and (2) Without using equipment that is CERTIFIED FOR THAT TYPE of appliance in 82.158.

(c)   **November 15, 1993** — No person may manufacture or import recycling or recovery equipment for use during the maintenance service or repair of appliances except MVACs and no person may manufacture or import recycling or recovery equipment for use during the disposal of appliances except small appliances unless the equipment is certified per 82.158(b),(d) or (f).

(e)   **August 12, 1993** — No person may open appliances except MVACs for maintenance, service or repair and no person may dispose of appliances except for small appliances, MVACs and MVAC-like appliances unless such person has certified to the administrator per 82.162 that such person has acquired certified recovery or recycling equipment and is complying with the applicable requirements of this subpart.

(f)   **August 12, 1993** — No person may recover refrigerant from small appliances, MVACs and MVAC-like appliances for purposes of disposal of the appliances unless such person certifies per 82.162 that such person has acquired recovery equipment that meets the standards in 82.158(l)/(m) as applicable and that such person is complying with the applicable requirements.

(g)   **August 12, 1993** — Until November 13, 1995, no person may sell or offer for sale for use as a refrigerant any class I or II substance consisting wholly or in part of used refrigerant unless the class I or II substance has been reclaimed as defined in 82.152(q). AND

(h)   Has been reclaimed by a person who has been certified as a reclaimer per 82.164

(i)   **August 12, 1993** — No person reclaiming refrigerant may release more than 1.5% of the refrigerant received by them.

(j)   **November 15, 1993** — No person may sell or distribute or offer for sale or distribution any appliances except small appliances unless such equipment is equipped with a servicing aperture to facilitate the removal of refrigerant at servicing and disposal.

(k)   **November 15, 1993** — No person may sell or distribute or offer for sale or distribution any small appliance unless such equipment is equipped with a process stub to facilitate the removal of refrigerant at servicing and disposal.

(l)   **November 15, 1993** — No person may open an appliance except for an MVAC and no person may dispose of an appliance except for a small appliance, MVAC or MVAC-like appliance, unless such person has been certified as a technician for that type of appliance per 82.161.

(n)   **November 14, 1994** — No person may sell or distribute or offer for sale or distribution any class I or II substance for use as a refrigerant to any person unless:

(1)   The buyer is certified as a Type I, II, III or universal technician per 82.161;
(2)   The buyer is certified per 40 CFR part 82, Subpart B, (section 609, 82.42);
(3)   The refrigerant is sold only for eventual resale to certified technicians or to appliance manufacturers (e.g., sold by a manufacturer to a wholesaler, sold by technician to a reclaimer);
(4)   The refrigerant is sold to an appliance manufacturer;
(5)   The refrigerant is contained in an appliance; or
(6)   The refrigerant is charged into an appliance by a certified technician during maintenance, service or repair.

(o)   IT IS A VIOLATION OF THIS SUBPART TO ACCEPT A SIGNED STATEMENT PER 82.156(f)(2) IF THE PERSON KNEW OR HAD REASON TO KNOW THAT SUCH A SIGNED STATEMENT IS FALSE.

SECTION 608, REQUIRED PRACTICES, 82.156, EFFECTIVE DATES

(a)   **July 13, 1993** — All persons: Opening appliances, except for MVACs, for maintenance, service or repair must evacuate the refrigerant in either the entire unit or the part to be serviced (if the latter can be isolated) to a system receiver or a recovery or recycling machine certified per 82.158 OR: Disposing of appliances except for small appliances, MVACs and MVAC-like appliances, must evacuate the refrigerant in the entire unit to a recovery or recycling machine certified per 82.158. AND

(1)   Opening appliances except for small appliances, MVACs and MVAC-like appliances for maintenance, service or repair must evacuate to the levels in Table 1 before opening the appliance, unless;

(i)   Evacuation of the appliance to the atmosphere is not to be performed after completion of the maintenance, service or repair and the maintenance service or repair is not major as defined at 82.152(j); or

(ii)   Due to the leaks in the appliance evacuation to the levels in Table 1 is not attainable, or would substantially contaminate the refrigerant being recovered. IN ALL CASES 82.156(a)(2) MUST BE FOLLOWED.

MORE DATES TO REMEMBER

1985

Vienna Convention, international negotiators to draw up a plan for worldwide action on the CFC issue.

September, 1987

The United States signed the Montreal Protocol on substances that deplete the ozone layer. The Protocol called for the production and consumption of CFCs and Halon to be frozen at 1986 levels beginning July 1, 1989 and January 1, 1992 respectively, and for the CFCs to be reduced to 50 percent of 1986 levels by 1998.

August 12, 1988

EPA gives regulations to implement the 1987 Protocol thru a system of tradable allowances for manufacturers based on their 1986 production levels, then reducing the allowances according to the schedule specified in the Protocol.

1989

Omnibus Budget Reconciliation Act imposes TAX on the sale of virgin CFCs and other chemicals that deplete the ozone layer, with exemptions for exports and recycling.

June, 1990

Protocol amended for CFCs and Halons to be phased out by January 1, 2000. HCFCs to be phased out by 2020 if possible and no later than 2040.

January 1, 1992

MVAC shops must have approved recovery/recycling equipment and technicians must be trained to properly use it.

1991

Excise tax is amended to include methylchloroform, carbon tetrachloride and other CFCs regulated by the amended Montreal Protocol and Title VI of the Clean Air Act.

July 1, 1992

Prohibition on venting during servicing, repair and disposal of an appliance or industrial process refrigeration or to knowingly vent or otherwise knowingly release or dispose of class I and class II refrigerants that permits these substances to enter the environment. FR Vol 58, No 92, 5/14/93.

July 14, 1992

EPA final Section 609 regulations are published.

August 13, 1992

Effective this date no person repairing or servicing motor vehicles for consideration may perform ANY service on a motor vehicle air conditioner (1) without using approved equipment approved pursuant to 82.36 and (2) unless such person has been properly trained and certified by a technician certification program approved by the administrator pursuant to

82.40. (Section 609 prohibitions 82.34). May be extended to January 1, 1993 if servicing less than 100 vehicles, see 82.42(a)(2). SEE January 1, 1993.

November 15, 1992

Prohibition on selling, distributing or offering for sale or distribution any class I or II substance as a refrigerant in MVACs in containers smaller than 20 lbs, unless the purchaser is properly trained and certified under 82.40 or certifies the containers for resale under 82.42(b)(4).

January 1, 1993

No later than this date anyone repairing or servicing MVACs shall certify that they have acquired and are properly using approved equipment and all individuals authorized to use the equipment are properly trained and certified. (Section 609, 82.42). SEE also August 13, 1992.

February 11, 1993

Substances designated as class I or class II must have their containers and products marked per 40 CFR PART 82, Subpart E-Labeling. Section 611 FR 2/11/93, page 8164.

May 15, 1993

As of this date any substance designated as a class I or class II substance AFTER February 11, 1993 must be labeled per 40 CFR part 82, Subpart E - The Labeling of Products Using Ozone-Depleting Substances. Section 611 FR 2/11/93 page 8164 -8169.

June 14, 1993

Refrigerant recycling program established for ozone depleting refrigerants recovered during the servicing and disposal of air conditioning or refrigeration equipment by Section 608. This also covers MVAC that are being disposed of. Section 609 covers other 3R requirements for MVAC servicing. FR Vol 58, No. 92, 5/14/93, 28664, 28666.

July 13, 1993

Landfill operators, scrap recyclers or anyone else who accepts for disposal, small appliances, room air conditioning, MVACs or MVAC-like appliances MUST either: recover any remaining refrigerant OR verify that refrigerant has been removed and possess a signed statement from the person whom the appliance is obtained that they removed all refrigerants and the statement must have the name and address of the person who recovered the refrigerant, the date recovered or certify that it will be removed before delivery.

Operators/recyclers must post notices or warning signs or letters to suppliers or other equivalent means.

August 12, 1993

From this date until May 15, 1995 used refrigerants must be reclaimed to 608 and/or 609 standards before being sold as refrigerants. Reclaimers must be certified and not release more that 1.5% of refrigerant received. See 82.36 and Appendix A of Section 609 and 82.158 and 82.154 of section 608.

November 22, 1993

Federal agencies must modify their procurement regulations to comply with section 613.

November 14, 1994

MANDATORY CERTIFICATION DATE for Section 608; if you are not certified by this date DO NOT WORK ON APPLIANCES COVERED BY SECTION 608. See 82.154(l)(n).

November 15, 1993

Certification date for recovery/recycling equipment. See 82.158(b((c)(d)(e)(g).

November 15, 1995

Definition of refrigerants for MVAC as type I and II substances shall be amended to include any substitute substance. (82.32 (f).

January 1, 1996

CFCs, carbon tetrachloride, and methylchloroform phased out. FR 4/12/93 page 19082.

SECTION 608 EVACUATION LEVELS

Section 608 EVACUATION LEVELS for appliances except small appliances, MVACs and MVAC-like appliances. From tables 1, 2 and 3 of Section 608.

There are two standards to be met, determined by the date of manufacture of your recovery or recycling equipment. November 15, 1993 is the date for old/new specs for vacuum.

APPLIANCE EVACUATION LEVELS
Except Small Alliances, MVACs and MVAC-Like APPliances

	HG in.
HCFC-22 appliance, LESS than 200 lbs refrigerant	0/0
HCFC-22 appliance, MORE than 200 lbs refrigerant	4/10
High pressure appliances, LESS than 200 lbs refrigerant	4/10
High pressure appliances, MORE than 200 lbs refrigerant	4/15
Very high pressure appliances	0/0
Low-pressure appliances	25/25

Helpful Hints

↪ Notice that the first four charges are LESS, MORE, LESS, MORE.
↪ The first two are for HCFC-22 appliances, then two high pressure appliances.
↪ Reconstruct the chart for the test.

    The first vacuum is 0/0 for LESS than and HCFC-22.
    The second vacuum is 4/10 for MORE than and HCFC-22.
    The third vacuum is 4/10 for LESS than and HIGH pressure.
    The fourth vacuum is 4/15 for MORE than and HIGH pressure.
    The fifth vacuum is 0/0 for VERY high; notice there was no change in vacuum levels.
    The sixth is 25/25 for LOW pressure and again no change in level.

↪ AFTER the test begins write the following chart on scratch paper:

    22-0/0
    22+4/10
    HP-4/10
    HP+4/15
    VHP0/0
    LP25/25

CHAPTER FOUR

CERTIFICATION

Effective June 14, 1993, NO person may knowingly vent class I or II refrigerants into the atmosphere. Furthermore, the required practices of CFR 82-156 and recovery or recycling machines must be used that meet the requirements of CFR 82-158. *ANYONE* following these requirements may work on air conditioning equipment *UNTIL* November 14, 1994. On November 14, 1994 *NO PERSON* may open an appliance or dispose of an appliance except for small appliances, MVACs or MVAC-like appliances unless such person HAS BEEN CERTIFIED as a technician for that TYPE of appliance as specified in CFR 82-161.

For example: If you are certified and supervising a three man work crew, none of whom are certified, and they are charging air conditioning systems in a housing sub division, are any EPA regulations being violated? No, as long as CFR 82-156 & 158 are followed and the date of the service is before November 14, 1994.

There are technicians that will argue that without EPA certification, they may still work on some electrical controls, replace fan motors, belts, filters and fuses, if they are not part of the refrigerant system, and will not cause venting or a release of refrigerant into the environment. The three questions I would ask are: Can my actions reasonably be expected to cause release of refrigerant from the system? If the answer is yes, then you MUST have proper EPA certification and recovery equipment. How will I justify to an EPA inspector that the release could not have been reasonably expected to happen? Do I know this system well enough to risk a fine? It's best not to take chances, accidents do happen, and if you should cause CFC venting you may incur heavy fines.

There are five EPA certifications available, all of which allow the technician to purchase any refrigerant. However, the type of certification that you have LIMITS the equipment that you may service, maintain or repair. The EPA certification not only allows the purchase of refrigerants after November 14, 1994 but by design of the program also specifies the types of equipment you may service. The various certification types and what each may work on are identified below.

TYPE I CERTIFICATION

A TYPE I technician can work on any SMALL appliance that was fully manufactured, charged, and hermetically sealed at the factory, with five (5) pounds or less of refrigeration. Home refrigerators, freezers, window air conditioners, packaged terminal air conditioners/terminal heat pumps, dehumidifiers, vending machines, water coolers, under the counter ice makers, and drinking water coolers are listed. The examination may be of the mail-in format with a score of 84% minimum for passing. A proctored closed book examination requires a passing score of 70% or better.

TYPE I technicians — servicing ONLY small appliances — can use recovery equipment that is EITHER the active or "self contained" type or the passive "system dependent" type. Self contained recovery equipment provides a method of transferring refrigerants OUT of the appliance and into the recovery vessel without assistance from the appliance.

System dependent recovery equipment relies on the appliance compressor to pump the refrigerant out, and/or on the pressure in the system to assist in recovery. There is no other external mechanism used for evacuation. Chilling the recovery vessel can decrease the time required for type I recovery, and remember that refrigerant migrates to the coldest point. System dependent recovery equipment would be connected to the small appliance compressor DISCHARGE line if the compressor is operable. If the compressor is inoperative then recovery is made thru *BOTH* high and low side connections.

The recovery equipment that your shop uses must be approved and registered with the EPA and the recovery unit must be used as the EPA service procedures recommend. If the manufacturers instructions conflict with the EPA procedures, consult with the EPA and manufacturer to resolve the conflict.

Small appliance recovery requirements

Effective July 13, 1993, ALL persons opening appliances except for MVACs for maintenance, service, or repair must evacuate the refrigerant in either the entire unit or the part to be serviced to a system receiver or a recovery/recycling machine certified by CFR 82.158(d)(e).

Equipment manufactured before November 15, 1993 must recover 80% of the refrigerant whether the compressor operates or not, or evacuate the small appliance to 4 inches of vacuum per CFR 82-158(e). On or after November 15, 1993 the equipment must recover 90% of the refrigerant *IF* the compressor operates or 80% if the compressor is inoperative, or evacuate the small appliance to four (4) inches of mercury vacuum per CFR 82.156(a)(4) and (c)(d). If you are disposing of a small appliance the requirements are: recover 90% if the compressor operates, 80% if the compressor is inoperative or evacuate the small appliance to four inches of mercury vacuum per CFR 82.156(h). Access may be by clamp on piercing or line tap valves which should be leak tested before continuing the work. These valve types may eventually leak. They should be removed after servicing.

Small appliance refrigerant leaks

Unlike other appliances, there currently isn't a requirement to repair small leaking appliances under federal EPA guidelines. However, states MAY require the repair of leaks in small appliances. CFC-12 and HCFC-22 are predominately used for the refrigerants in these appliances. There are currently no drop-in replacements for CFC-12 or HCFC-22. HFC-134A is being used as a substitute for CFC-12, BUT requires replacement of rubber seals, system lubricant, desiccant and other equipment modifications. The appliance manufacturer should always be consulted for feasibility of appliance conversion to an alternative refrigerant.

TYPE II CERTIFICATION

A TYPE II technician can work on and dispose of high or very high pressure appliances and MVAC-like appliances, EXCEPT small appliances and MVACs. A high pressure appliance uses a refrigerant with a boiling point between -50° and 10°c at atmospheric pressure (29.9 inches of mercury). Refrigerants 12, 22, 114, 500, 502 are typically used in high pressure systems. A *very high* pressure appliance uses refrigerants with a boiling point BELOW -50°c at atmospheric pressure. Refrigerants 13, 503 and others may be used. MVAC-like appliance means mechanical vapor compression, open drive compressor appliances used to cool the drivers or passengers compartment of a NON-road motor vehicle. Agricultural and construction vehicles are examples. *REMEMBER* that to work on motor vehicle air conditioners you *MUST* be EPA certified by section 609. Recovery equipment used with MVAC and MVAC-like appliances must meet the requirements of section 609, CFR 82.36(a). MVAC-like appliance recovery equipment standards are referred to CFR 82.36(a) by section 608, CFR 82-158(a).

A TYPE II technician may work on the following equipment covered by SECTION 608:

RESIDENTIAL AIR CONDITIONING — includes window units, packaged terminal air conditioners, central air conditioners, light commercial conditioners and heat pumps which rely on HCFC-22.

TRANSPORT REFRIGERATION — consists of refrigerated ship holds, truck trailers, railway freight cars and other shipping containers. Ship holds use HCFC-22 and ammonia — others use CFC-12.

COMMERCIAL REFRIGERATION — covers retail food and cold storage warehouses. RETAIL FOOD equipment includes small reach-in refrigerators and freezers, refrigerated display cases, walk-in coolers and freezers and large parallel systems. Refrigerants are CFC-12, 502 and HCFC-22. COLD STORAGE warehouses are used to store meat, produce, dairy products and other perishable goods.

COMMERCIAL COMFORT AIR CONDITIONING — utilizes chillers for temperature and humidity regulation in offices, hotels, shopping centers and other large buildings. The three types of chillers are named after the type compressor used. They are the centrifugal, reciprocating and the screw compressor. Refrigerants are CFC-12, -500, and HCFC-22 for centrifugal, CFC-12 and HCFC-22 for reciprocating and HCFC-22 for screw chillers.

INDUSTRIAL PROCESS REFRIGERATION — includes ice machines, ice rinks and customized systems used in the chemical, pharmaceutical, petrochemical and manufacturing industries. Refrigerants are CFC-12, -500, -502 and HCFC-22, with refrigerant charges as high as 20,000 lbs for some centrifugal units!

MVAC-LIKE APPLIANCES — includes farm equipment, airplanes, construction equipment and off-road equipment, boats, and the disposal of MVACs.

NOTE: Chillers using CFC-11, 113 and HCFC-123 are LOW PRESSURE chillers and are NOT serviced by Type II technicians. Type III technicians service LOW Pressure appliances.

High pressure appliance recovery requirements

Table 1 in section 608 lists the evacuation levels required by charge, pressure and one refrigerant type. Table 1 is NOT for small appliances. CFR 82-156(a) thru (g) need to be read closely for exceptions to Table 1. It's important to remember that even the exceptions require evacuation to 0 psig as a minimum.

You are *REQUIRED* by CFR 82-156 (b) to have at least one piece of certified, SELF-CONTAINED recovery equipment available at your place of business. *SYSTEM-DEPENDENT* equipment is restricted by CFR 82-156(c), to appliances normally containing 15 lbs of refrigerant or less. Refrigerant may be returned to the appliance from which it is recovered or to another appliance *OWNED* by the same person without recycling or reclaiming, unless the appliance is an MVAC-like appliance per CFR 82-156(e). Reuse of refrigerant removed from MVAC and MVAC-like appliances is covered in section 609, appendix A, 1.

Scope

Type II technicians may work on MVAC-LIKE appliances and dispose of MVAC and MVAC-like appliances per CFR 82-150(b), 82-154(a)(2)(b), 82-156(f)(g) and/or section 609 82-30(b). The recovery equipment standards for Type II servicing are listed in CFR 82-158.

CFR 82-156 lists the required practices for Type II technicians. You should become familiar with them, especially the required evacuation levels for refrigerants and charges.

Leaks

1. Owners of commercial refrigeration and industrial process refrigeration equipment MUST have all leaks repaired if the equipment is leaking at a rate such that the loss of refrigerant will exceed 35% of the total charge during a 12 month period, CFR 82-156(i)(1)(4).

2. Owners of appliances normally containing MORE than 50 lbs of refrigerant and not covered by 1. must have all leaks repaired if the appliance is leaking at a rate such that the loss of refrigerant will exceed 15% of the total charge during a 12 month period, CFR 82-156(i)(2)(4).

> **EXCEPTION to 1 and 2** — Repair of leaks is not required IF, within 30 days, owners develop a one-year retrofit or retirement plan for the leaking equipment. The plan (or copy) must be kept at the equipment site. The original must be available for EPA inspection on request. The plan must be dated *AND ALL WORK MUST* be completed within one year of the plans date, CFR 81-156(i)(3).

3. Leak repair must be made within 30 days of discovery or within 30 days of when the leak(s) should have been discovered if the owners intentionally shielded themselves from information which would have revealed a leak, CFR 82-156(i)(4).

4. Records — technicians *MUST* provide the owners of equipment holding more than 50 lbs of refrigerant charge with an invoice or other documentation indicating how much refrigerant was added to the equipment. Service technicians working on systems for the first time should request the equipment service records to review prior service, refrigerant use, and charging history. If recharging has occurred, evaluate the leak rates as specified in number 1 and 2. ANYTIME an appliance requires refrigerant, the servicing technician must provide the owner with a service receipt listing the amount of refrigerant added. The servicing technician's company should also send a letter to the owner reconfirming that refrigerant was added. This letter should also notify the company that owns the system whether or not the allowable leak rates were exceeded, that repair is required and the time period allowed by EPA for repairs. This procedure protects the company if the appliance owner tries to avoid repairs or claims the service technician did not inform them of leaks or refrigerant charging. It is also worth $2.00 to *SEND THIS LETTER CERTIFIED MAIL* to protect your business and have a signed receipt for your files.

Developing leak abatement plans for clients can be profitable. You may be able to design a retrofit or replacement system for your clients. If the appliance is leaking and meets or exceeds the specified rate of leakage YOU MUST tell the appliance owner of the leak rate and that the appliance must be repaired within 30 days.

TYPE III CERTIFICATION

A TYPE III technician can work on and dispose of LOW-pressure appliances. Low-pressure appliances use refrigerants with a boiling point above 10°c at atmospheric pressure. Refrigerants 11, 113, 123 and others may be used.

The equipments are CHILLERS, the screw, centrifugal and reciprocating compressors. Low pressure also means the equipment operates below atmospheric pressure or in a vacuum. Air is pulled in so purge units are used to remove the air and other contaminates. A purge unit pulls gasses off the top of the chiller condenser by compressor suction. Then the compressor compresses and pumps the gasses into the purge drum where they are condensed and separated.

WARNING — Low pressure systems have a RUPTURE DISK that will open at 15 lbs pressure so be careful when charging. Refrigerant will escape to atmosphere from the evaporator when this disc operates. You can also blow it with nitrogen gas and with the recovery machine high pressure cut-out switch set above 10psig. If you purchased recovery equipment just for low pressure equipment it may not be suitable for use with high pressure appliances because of the liquid pump design. Your recovery vessel may also have a 15 psig rupture disk.

RECOVERY

The last line of table one's evacuation requirements for low pressure appliances is easy to confuse. Evacuation requirements are not the same values. The before value is 25in.Hg gauge and the on/after date of November, 15, 1993, value is 25mm Hg absolute which equals 29 in. Hg gage. Before starting recovery verify the high pressure cut out setting on the recovery equipment (10psig). Start with liquid recovery first then vapor recovery.

UNIVERSAL CERTIFICATION

A UNIVERSAL technician may work on all of the equipments that types I, II and III are permitted to work on. Universal technicians must be certified under section 609 to work on motor vehicle air conditioners.

SECTION 609 CERTIFICATION

A section 609 certification permits technicians to work on and dispose of motor vehicle air conditioners, CFR 82.30 and MVAC like appliances CFR 82.161(a)(5). Refrigerants used are CFC 12 and HFC 134A. The certification test may be closed book or open book depending on the testing organizations program.

Recovery equipment requirements are in CFR 82-36 and Appendix A to subpart B-Standard for Recycle/Recover Equipment. Recovered refrigerant may not be returned to the appliance unless the requirements of Appendix A are met.

For many of those seeking certification, the 608 type I certification is all that is needed. A listing of testing organization's is printed in Appendix II of this book. Call the EPA Stratospheric Ozone Hotline (800-296-1996) for an updated list of names and phone numbers of testing organizations. The EPA hot line will not have the testing dates available. You must contact the testing organizations individually to obtain a copy of their testing schedule. The list in this book is current up to the date of printing. The EPA may add new testing organizations or possibly remove testing organizations from the approved list that have not passed an EPA testing site evaluation.

CHAPTER FIVE

FREON (3R)
RECOVERY, RECYCLING, & RECLAIMING

The type of refrigeration equipment worked on determines — as a minimum — the type of refrigerant recovery equipment required by the EPA. Your refrigerant recovery equipment must be certified by and registered with EPA. Employees who use recovery equipment must be trained in the proper use of the recovery equipment and that training must be documented. EPA specifies what recovery equipment must be used by the refrigeration technician just as the technician's certification type specifies the appliances that may be worked on. Listing your recovery/recycling requirements and your methods of servicing appliances along with the refrigerants recovered, will enable the technician/business owner to make recovery equipment purchase selections that provide the most servicing flexibility in relation to the equipment cost.

TYPE I appliance servicing technicians may have either self-contained recovery equipment or system dependent recovery equipment. TYPE II and TYPE III appliances require the technician or business to have at least one piece of certified, self-contained recovery equipment, reference CFR 82-156(b). You may also have system-dependent recovery equipment but 82-156(c) restricts usage to equipment containing 15 lbs. or less of refrigerant. Type III appliances, low pressure — usually having many pounds of liquid to recover — pose special servicing considerations. The quantity of liquid and pressures encountered in a number of Type III low pressure systems would in some applications require days to recover the refrigerant using vapor recovery. A high capacity liquid pump is required along with LARGE HOSES to reduce liquid flow restrictions with high volume systems. Large diameter pipes could be used for increasing flow. As a minimum, use hoses of greater than 1/4 in for recovery, such as 3/8in. The larger diameter speeds up refrigerant vapor recovery. Also the recovery vessel must have a rupture disk set at 15 psi. The recovery unit must have the high pressure cut-out switch set to operate at 10 psi. — this is not adjustable. If you have machines that look similar but have different high pressure settings make sure the machine differences are conspicuously marked to prevent possible damage to the low pressure appliance and to prevent breaching of the recovery vessel rupture disk. The recovery vessel needs to be checked BEFORE starting the recovery process to insure that the vessel's rupture disk pressure setting is compatible with the recovery equipments high pressure control setting. The universal technician must use the recovery equipment specified for the type of appliance being serviced.

MVAC and MVAC-like appliances recovery requirements are listed in 82-158(f)(g)(l)and 82-36(a) and Appendix A to Subpart B - Standard for recycle/recover equipments.

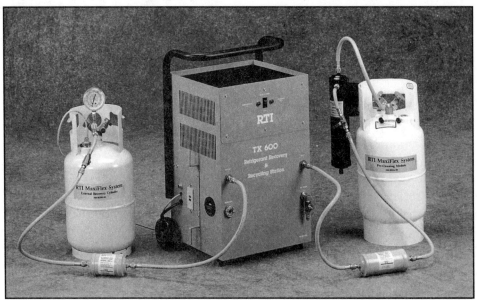

RTI MaxiFlex System - Recovery/Recycling/Refining System
Complements of RTI Technologies, Inc. (800-469-2321)

THE 3Rs
HOW THEY DIFFER

RECOVERY provides a method to remove and store refrigerant from an appliance without testing or processing. The process of extracting and cleaning refrigerant that is removed from an appliance for reuse — without meeting all of the requirements for reclamation — is the *RECYCLE* method. Freon is cleaned by using the oil separation method and by processing the liquid through single and multiple replaceable core filter-dryers that reduce moisture, acidity and particulates. The cleaning process normally takes place on the job site or at the local service site. Non condensable gasses and air are also removed. The reprocessing of refrigerants to new product specifications, specifically ARI Standard 700-1988,is called *RECLAIMING* and usually is done at a reprocessing or manufacturing facility often by distillation. Chemical analysis is required to verify refrigerant purity.

The reuse of refrigerants is allowed under section 608 as long as the refrigerant is returned to the same equipment or other equipment owned by the same person or company; otherwise it must be reclaimed. Prior to reusing refrigerant check for non-condensables. Testing for non-condensable gasses of a known refrigerant may be accomplished using the procedure listed in Section 609 Appendix A, Recommended Service Procedures, available from the EPA hotline. A standard temperature-pressure chart should be used rather than Table 1 of Appendix A, since it is for R-12 only.

Before reusing recovered refrigerant or recycling refrigerant for recharging the system it was removed from, determine what type of warranty you will give it. Few technicians have a laboratory in their hip pocket that can certify that the refrigerant removed met ARI Standard 700-1988 or that the same refrigerant removed and returned meets the ARI Standard. Companies need to have a policy on the reuse of the owners' refrigerant, and it needs to be discussed with the owner *BEFORE* service that would require recovery of refrigerant. You must also verify if your state, county or city have laws that regulate the use of recycled or reclaimed refrigerants. Another concern is whether equipment warranties will be voided by reusing refrigerant.

Section 609 does not allow for reuse of refrigerant unless it is recycled and meets the requirements of Appendix A to Subpart B-Standard for Recycle/Recover equipment. Some automotive servicing equipment will recycle R-12 refrigerant and recharge it to the system it came from plus some additional new refrigerant to bring the MVAC to rated capacity. As long as the equipment is marked with a label that attests to design certification meeting Society of Automotive Engineers (SAE) J1991 Standard, the equipment meets EPA recycling requirements for R-12 REFRIGERANT CLEANING for reuse per Appendix A to Subpart B-Standard for Recycle/Recovery Equipment. Section 608, 82-154(g) prohibits selling used refrigerant unless it has been reclaimed.

All 3R equipment has some degree of filtering of vapor or liquid to protect the 3R equipment. If for some reason the filter is missing from your recovery unit or it did not have one when purchased, install a filter on the suction line now! Recycling equipment has more levels of contaminate removal than a recovery unit. Any reclaimed refrigerant that is sold must be tested to verify that it meets ARI Standard 700.

The required levels of evacuation for all recovery equipment certified by Section 608 are listed in table 1 of Section 608 with before and after dates of manufacture. Recovery efficiency equals the total refrigerant recovered divided by the total refrigerant recoverable, times 100%. Refrigerant recoverable would be the appliance charge of refrigerant specified on the manufacturer's data plate. Section 609 Appendix A to Subpart B, 7.1, lists the standard of recovery as reducing the system pressure below atmospheric to a minimum of 102MM of mercury or 26in of HG vacuum. Section 609 also lists Society of Automotive Engineers (SAE) standards for R-12 recovery and recycling equipment. J1990 is the standard for recovery and recycling equipment components such as filtration, connectors on hoses and hose specifications for R-12 processing. J2210 is for recovery and recycling of R-134A, but is not listed in section 609. Appendix A to Subpart B. 3. lists the purity specifications for R-12, 3.1 Moisture 15ppm by weight, 3.2 Refrigerant oil 4000ppm by weight, 3.3 Noncondensable gasses (air) 330ppm by weight.

Other SAE Standards are, J1991 purity standard for R-12, J2099 purity standard for R-134a, J2211 recovery/recycle of R-134a in automotive appliances, J1989 recovery and recycling of R-12 in automotive appliances, J2196 R-12 and R-134a hose markings. Remember how helium escapes from a balloon? Helium has molecules so small that it can slip past the rubber molecules in the balloon. R-134a has molecules smaller than R-12 so service hoses for R-134a have to be made from denser compounds. R-134a hoses have quick disconnect fittings (no threads) on equipment fittings. Service hoses must have shutoff devices located within 30cm (12in) of the connection point, Appendix A to Subpart B(7.6).

SECTION 608 RECOVERY

Anytime you are going to do recovery work ensure that the appliance has a charge to recover. Otherwise you may recover air into your vessel and have to purge your system. Many Type II and Type III appliances have complex valving for isolation of major components for servicing. Closely inspect systems to ensure there are no pockets of refrigerant ready to spew out to the atmosphere.

Vapor is recovered with self-contained equipment when the recovery units compressor lowers the pressure in the appliance and the refrigerant boils off producing more vapor. The discharge is condensed to a liquid and stored in the recovery vessel.

CAUTION

If your recovery/recycling equipment's compressor is cooled by the flow of incoming vapor check for an auto shut off function when it reaches the required vacuum. If not, keep an eye on the vacuum and manually shut the unit off when recovery levels are met. By not doing this, you can starve the compressor of cooling vapor that can result in a burned out recovery unit. Chilling the recovery vessel can help speed up recovery except when recovery is by the push/pull method.

Table 1 on page 131 covers evacuation levels for appliances except small appliances and MVAC and MVAC-like appliances. For small appliances remove 90% of refrigerant if the compressor works, 80% if the compressor is inoperative or evacuate to 4in of HG.

A refrigerant recovery vessel is marked by a yellow top with a grey body. They are DOT or UL approved for use as recovery vessels. **DISPOSABLE SERVICE TANKS OR JUGS (DOT-39) ARE NOT APPROVED FOR RECOVERY. IT IS A VIOLATION OF LAW TO REFILL DISPOSABLES WITH ANYTHING BUT A VACUUM.** The right thing to do is *ALWAYS EVACUATE* the jug first, then drill or punch a hole in it before disposing of it. Some new refrigerants, when mixed with air, can *EXPLODE* when exposed to an ignition source such as a drill or sparks from a punch. 49CFR, Section 173.304 covers DOT regulations on cylinders if you want to know more.

NOTE

It is a violation of 49CFR to transport disposable vessels of refrigerant in your service truck unless the jug is in its cardboard shipping container! One more thing on disposables, resist the temptation to reuse them as portable air supplies in the shop or at job sites. OSHA fines can be very costly.

All hoses must have shutoff valves or low loss fittings which must close automatically or be closed manually. They should be short to reduce excessive venting. Some recovery machines will even evacuate the hoses. Use hoses at least as large as the fittings they will attach to; this speeds up recovery time.

Push/Pull Recovery (differential pressure) Method is defined as the process of transferring liquid refrigerant from a refrigeration system to a receiving vessel that has both liquid and vapor valves, by lowering the pressure in the vessel and raising the pressure in the system and by connecting a separate line between the system liquid port and the receiving vessel. This method is used with large systems where it would "take forever" to recover using a compressor recovery machine. Compressor recovery machines when used with liquids must have an accumulator or some liquid separation method to protect the compressor. Consider a recovery machine that has a liquid pump for fast liquid recovery if the push/pull method is not practical for what you service. If you frequently work on high capacity (hundreds of pounds of refrigerant) systems, push/pull is probably more economical and efficient for you.

Push/Pull equipment connections

The disabled unit's liquid side is connected to the recovery vessel liquid valve. The recovery vessel's vapor valve is connected to the recovery machine vapor inlet port. A filter dryer should always be in this line to protect your machine. The recovery machine outlet port would connect to the vapor side of the disabled unit. Energizing the recovery unit lowers the pressure in the recovery vessel and increases the pressure in the disabled unit, pushing and pulling the liquid into the recovery vessel. A standard self-contained recovery unit can accomplish this process. This method can save you the price of a liquid pump recovery unit that may be limited to liquids only.

GENERAL INFORMATION

Required practices for opening appliances except MVACs are covered in section 608, 82-156. For MVACs Section 609, Appendix A to Subpart B-Standard for Recycle/Recover Equipment, Recommended Service Procedure for the containment of R-12, lists the required practices. Connecting a set of gauges to any appliance constitutes OPENING that appliance. If the manufacturer's instructions for recovery equipment operation conflict with the EPA required practices FOLLOW THE EPA REQUIREMENTS. Recovery machines are built to recover class I and class II refrigerants and HCF refrigerants. Recovery equipment use with "other" refrigerants or gasses, SHOULD NEVER BE DONE, will likely void your warranty and severe damage to your equipment may result.

Disposal of replacement parts may be regulated by state and local laws, filters, desiccants, etc because of contaminates they may contain. Proper disposal procedures of those parts should be in your training plan for recovery equipment operation. Recovery equipment is costly; as a minimum follow the manufacturer's service interval recommendations. Most people change the oil and filter on their car BEFORE the recommended interval to increase the life of their car and keep wear to a minimum. You can do the same thing with your recovery equipment by servicing it more frequently than required by the manufacturer and NEVER exceeding manufacturers service intervals.

Check with your wholesale house, the manufacturer of your recovery/recycling equipment and refrigerant manufacturers for reclaiming services available to you or your business. If the appliance is under warranty, will it be voided by using reclaimed or recycled refrigerant? Do not fill recovery vessels more than 80% full of liquid, 49CFR, 173.304; this is NOT an EPA requirement but a Department of Transportation (DOT) requirement.

HVAC&R Commercial & Automotive Recovery/Recycling Equipment
Complements of RTI Technologies, Inc (800-468-2321)

CHAPTER SIX

SECTION 608, 609 & 611
GROUP I PRACTICE QUESTIONS

The practice questions are presented in several formats to help you retain the information. Select the most correct answer for each question. The sample question, number (1A), is an example of a multiple choice format and the correct answer is d. All practice question answers are located in Appendix 1.

EXAMPLE QUESTION (1A)

Which of the following statements about the earth is correct?
      a.   The earth is round and hollow
      b.   The earth doesn't rotate around the sun
      c.   The earth is the smallest planet in our solar system
      d.   The earth is round and rotates around the sun
(Our Solar System, Sentinel Publications, 1993)

Some questions have a reference identified in parentheses directly below the question. If you can't answer the question, go to the reference before looking in Appendix 1 for the answer. If a reference isn't specified, look through the Index and Table of Contents to locate the subject. If you need to brush up on air conditioning and/or refrigeration theory or system operation refer to references such as **The Refrigeration Service Engineers Society's (RSES)** *Service Application Manual* (SAM). This four volume set may be available at technical libraries or you can order it from RSES, 1666 Rand Rd., Des Plaines, IL 60016-3552, phone (708) 297-6464. Business News Publishing Company, Technical Book Division, P.O. Box 2600, Troy, MI 48007-2600, phone 800-837-1037, also publishes HVAC/R references including *Refrigeration Fundamentals* by William Gorman or *Basic Refrigeration* by Guy King.

It's also important to insure that your shop has the latest EPA updates for rule changes and final decisions. Stay in contact with the EPA through the many resources listed in this book and use their electronic bulletin board and toll free numbers to obtain the latest available information.

TEST TAKING STRATEGIES

The following strategies will help you improve your grades. Use these strategies on the practice tests in this book, on the closed book practice exam in Appendix III, and when you take your actual EPA certification exam. If you practice these techniques now, when you take the certification exam they will become second nature.

- ✎ Eliminate the answers in multiple choice questions that make no sense at all. You can often eliminate half of the answers through this method. If you have to guess an answer, you improve your chances through the process of elimination.
- ✎ Be skeptical when an answer includes words like, always, never, all, none, generally, or only. These words can be a trap. Only select an answer with these words in it if you are absolutely sure it is the right answer.
- ✎ If two answers have opposite meanings, look closer. Many times one of the two is correct.
- ✎ Place a mark next to answers that you are unsure about. After completing the remainder of the exam, go back and review these questions and make a final selection. Often, other questions that you've answered will jog your memory.
- ✎ One word can dramatically change the meaning of a sentence. Read each question word-for-word before answering.
- ✎ Don't let the test get the best of you. Build your confidence by answering the questions you know first. If the first question you read stumps you, skip it and go on to the next one. When you've completed most of the exam you can go back - if time permits - to the questions that you couldn't answer.
- ✎ Get plenty of rest the night before the exam.

STUDY HABITS

It's often helpful to study with a partner, someone to read the question and check your responses. It can be a fellow worker, a spouse, or just a good friend. If you don't have a partner, try recording the questions. Pause between each question so that you will have sufficient time to respond before the next question is read.

You can also record sections of the regulations and play them back during your daily commute to and from work. The more drill the better. The key to learning and retention is finding what works best for you. Try various study routines until you hit a combination that works. Try studying in 20 to 30 minute sessions with 5 minute breaks in between, or stretch it out to hour intervals. A good study routine will improve your retention and test scores.

Finally, this book should become a habit until you thoroughly understand the Clean Air Act and have successfully passed the certification exam. Try to devote at least one hour each day to study, and for heavens sake don't cheat yourself by writing the correct answers next to each question. The questions and answers were separated to help you retain and learn the material, not because I wanted to make it more difficult for you. The EPA questions don't come with answers; that's why some certification programs reportedly have a 40% failure rate.

The following questions were developed from the Federal Register and from official EPA material. If you can't answer a question, go to the stated reference, Index or Table of Contents to locate the subject. Study the appropriate reference material until you can answer the question. Take plenty of notes and use this book's margins to scratch down anything of significance that can help you through the exam, such as additional sources of information or cross references.

1. Ozone depletion may contribute to:
    a. Skin cancers   b. Cataracts
    c. Larger vegetables   d. a and b

2. September 1987 is the date when the
    a. Omnibus Budget Reconciliation Act taxed CFCs.
    b. Restriction on uses for CFCs was passed.
    c. Montreal Protocol was signed.
    d. Production limits on CFCs began.

3. De minimis servicing releases are violations if
    a. 40CFR part 82, Subpart B is followed.
    b. The release occurs before June 1, 1993.
    c. Practices in 40CFR part 82.156 are not used.
    d. More than 1.5% of the refrigerant is released.
(40CFR 82.156)
4. The restrictions on opening appliances, recovering refrigerant, disposal of appliances and selling used Class I or II refrigerants begins
    a. August 12, 1993   b. November 15, 1993
    c. June 14, 1993   d. July 13, 1993

5. A refrigerant with a boiling point of -50 degrees Centigrade at atmospheric pressure is used in
    a. Only appliances with low-loss fittings.
    b. Very high-pressure appliances.
    c. Only in industrial process refrigeration.
    d. Motor Vehicle Air Conditioners.
(40CFR 82.156)
6. The stratosphere begins and ends where in relation to sea level?
    a. 0 to 5 miles   b. 10 to 30 miles
    c. 0 to 6 miles   d. 6 to 30 miles

7. If you work on automotive air conditioners for free, absolutely no consideration for your work, are you required to allow an EPA inspector to see your records or allow access to your property?
    a. T   b. F

[CFR 82.42(b)(5) and 82.32(g)]
8. As a certified universal technician, you:
    a. May work on any type appliances.
    b. May work on MVAC systems.
    c. May purchase CFC-12 in sizes smaller than 20 lbs.
    d. May work with refrigerants after November 14, 1994.

9. Substitute substances must be recovered the same as CFC-12 effective
    a. 11/14/94   b. 11/15/93   c. 11/15/95   d. 7/13/92

10. Good ozone exists where?
    a. The Troposphere   b. The Ionosphere
    c. The Stratosphere   d. The Biosphere

11. When broken down by solar radiation the chemical element released by CFCs that damages ozone is
    a. Hydrogen   b. Oxygen
    c. Triatomic Oxygen   d. Chlorine

12. What causes ozone and oxygen molecules to be produced?
    a. Halons   b. photons   c. UV-A   d. UV-B

13. An increase in measured UV-B readings on the ground is an indication of what?
    a. Increased ozone in the stratosphere.
    b. Increased chlorine monoxide in the stratosphere.
    c. Increased tri-atomic oxygen in the troposphere.
    d. Global warming.

14. HCFC-22 has a higher ODP than CFC-12.
    a.T   b.F

15. A type I technician may work on appliances with
    a. 5 lbs. or less of refrigerant.
    b. Pitiot service stubs.
    c. 5 lbs. or more of refrigerant.
    d. 50 lbs. or more of refrigerant.

16. The restriction on purchasing CFC-12 in small cans started in
    a. September, 1987   b. November 15, 1992
    c. November 15, 1993   d. June, 1993

17. Which refrigerant does not contain chlorine?
    a.CFCs   b.HCFCs   c.HFCs.   d.HCFLCs.

18. Which refrigerant has an ODP of one?
    a. HFC-134A   b.CFC-12   c. HCFC-22   d.HCFLC-112

19. HCFC-22 is more unstable than CFC-12.
    a.T   b.F

20. The penalty for violations of section 608 are?
    a. Up to a $25,000 fine.
    b. Misdemeanors and up to $500 per day fine.
    c. Up to $25,000 per day, per occurrence.
    d. $2,500 maximum per day for all occurrences.

21. The bounty EPA will pay for turning in violations of the CAA is
    a. $25,000   b. $2,500   c. $1,000   d. $10,000

22. What does the term "drop in substitute" mean?
    a. A substitute refrigerant that does not require modification of existing appliance.
    b. A substitute refrigerant that requires some modification to the appliance.
    c. A substitute refrigerant like HFC-134A.
    d. A refrigerant that may be mixed with existing appliance refrigerant.

23. A refrigerant with an ozone depletion factor of 1 and an expected life of over 100 years is
    a. R-134A   b. R-22   c. R-12   d. none of these.

24. Agroscopic means
    a. acidity b. absorbs water c. alkaline d. base chlorides

25. Mineral oils are compatible with R-134A when used in
    a. low pressure equipment.   b. high pressure equipment.
    c. small appliances.   d. none of these.

26. What is the proper attire for working with R-22?
    a. Safety goggles and butyl-lined gloves.
    b. Safety goggles and leather gloves.
    c. Safety goggles.
    d. Safety goggles and latex gloves.

27. The liquid line ruptures in the equipment room where you are working. You should
    a. Run to the shut off valve and close off the leak.
    b. Turn off the compressor and wait for the air to clear.
    c. Leave the area and wait for the air to clear.
    d. Start immediate recovery to minimize loss.

28. Before removing hoses from service fittings you must
    a. Close the hose valves if not automatic type.
    b. Reach required evacuation level.
    c. Shut recovery vessel valve off.
    d. Turn appliance off.

29. A storage or recovery vessel must be
    a. UL or DOT approved.
    b. UL approved.
    c. Filled to rated capacity.
    d. Pressure tested every 4 years.

30. DOT regulations on transporting pressurized cylinders is covered in
    a. 29CFR part 1910.   b. 49CFR part 1910.
    c. 49CFR part 173.   d. 49CFR part 170.

31. Oxygen sensors are used
    a. in equipment rooms. b. in automobiles.
    c. in purge units.   d. in low pressure equipment.

32. As a minimum your DOT shipping papers must indicate what?
    a. Number of cylinders.
    b. Number of cylinders and weight.
    c. Number of cylinders, weight and warning label.
    d. Number of cylinders and type of refrigerants.

33. Sunlight can cause
    a. ozone.   b. cylinders to explode.
    c. chlorine monoxide.   d. all of the above.

34. From 31,680 feet to 158,400 feet above sea level scientists have measured increased levels of
    a. ozone.   b. chlorine monoxide.
    c. tri-atomic nitrogen. d. HFC-134A.

35. CFC-12 has been increasing in the upper atmosphere. Scientists believe that
    a. volcanic eruptions carried it there.
    b. supersonic aircraft are responsible for the increase.
    c. winds carry it aloft.
    d. tornados "suck" the CFC-12 into the stratosphere.

36. The prohibition on venting went into effect
    a. 6/1493. b. 5/14/93. c. 7/13/93. d. 11/15/93.

37. 608 technicians must use recovery equipment as of what date?
    a. 5/14/93. b. 6/14/93. c. 7/13/93. d. 8/12/93.

38. You comply with de-minimis release requirements if
    a. recovery equipment hoses have automatic shutoff connectors.
    b. recovery equipment and service gauges have automatic shutoff connectors.

c. all servicing hoses used have automatic shutoff connectors.

d. a and b.

39. Technician Type Universal may purchase all refrigerants
   a. except R-113.
   b. except HFC-134A.
   c. except R-12 in small containers.
   d. with no restrictions.

40. Self contained breathing apparatus
   a. is used by underwater divers.
   b. is used to prevent fainting.
   c. is used when charging HFC-134A systems.
   d. never used in confined spaces.

41. Refrigerant releases are de minimis if
   a. required practices are followed and approved recovery equipment is used.
   b. 82-156, 82-158 and 82.34 are followed.
   c. either (a) or (b) are followed.
   d. neither (a) nor (b) are followed.

42. This document called for the reduction of CFCs.
   a. Geneva convention.   b. Monrial Protocol.
   c. Brussels Protocol.   d. Montreal Protocol.

43. A small appliance must be evacuated to what level?
   a. 25mm Hg absolute.   b. 10 Hg vacuum.
   c. 0 Hg vacuum.   d. 4 Hg vacuum.

44 Junk yard Joe is properly using recovery equipment before disposing of small appliances. His recovery equipment is registered but Joe tells you that he is not type certified.
   a. Joe is in compliance.
   b. Joe is in violation of section 608.
   c. Turn Joe in for the bounty.
   d. (b) and (c)

45. Before a recovery vessel is used you should
   a. evacuate it to 30 Hg.
   b. check for DOT or UL approval.
   c. check hydrostatic test date.
   d. calculate maximum fill weight.

46. Failure to demonstrate proper recovery procedures may
   a. end in revocation of technicians Type certification.
   b. result in a citation.
   c. be acceptable.
   d. not matter.

47. Irregular heartbeat may be caused by
   a. excessive UV-B rays.
   b. exposure to refrigerant oils.
   c. ingesting contaminated refrigerant.
   d. excessive exposure to refrigerants.

48. HCFCs are not found in the stratosphere because
   a. they are heavier than air.
   b. they decompose at ground level.
   c. they wash out of the air when it rains.
   d. they decompose when exposed to nitrogen.

49. A material used to line gloves to protect the body from refrigerants is
   a. latex rubber b. neoprene c. butyl rubber d. cowhide.

50. The appliance group most responsible for CFC emissions is
   a. small appliances.
   b. purge units.
   c. chillers.
   d. MVAC and MVAC-like appliances.

51. The worst destroyer of ozone is
   a. HCFC-22.   b. R-123.   c. HFC-134A.   d. R-12.

52. ODP stands for
   a. oxygen depletion level.
   b. ozone depletion potential.
   c. ozone development plan.
   d. nothing.

53. Ozone depletion occurs primarily in
   a. the Arctic.   b. Antarctica.
   c. Aleutian islands.   d. the summer.

54. CFCs will not be manufactured for refrigerants as of
   a. 5/14/94. b. 1/1/95. c. 1/1/2000. d. 1/1/2002.

55. The only source of CFC-12 after production stops is
   a. recovery.   b. importing.   c. recycling.   d. disposal

56. Unfavorable chemical breakdown of refrigerants is usually caused by
   a. friction b. freon burn c. hydrofluoric acid d. water.

57. Lightly tapping the compressor and energizing heaters helps to
   a. release seized valves.
   b. release refrigerant from oil.
   c. speed up recovery.
   d. prevent acids from forming.

58. Polyolesters are
   a. contaminates of acids.   b. contaminates of water.
   c. synthetic oils.   d. refrigerant additives.

59. Unlike HFC refrigerants, CFC and HCFC refrigerants contain
   a. hydrogen b. oxygen c. hexaflouride d. chlorine

60. The uniform Mechanical Code prohibits
    a. smoking in mechanical rooms.
    b. oxygen sensors in mechanical rooms.
    c. stockpiling refrigerants in mechanical rooms.
    d. mixing of refrigerants.

61. Isolating chiller oil sumps
    a. helps collect all the oil.
    b. eliminates freon burn.
    c. reduces orifice plate damage.
    d. prevents refrigerant releases.

62. ASHRAE standard 15 requires that an equipment room using B1 refrigerant
    a. contain refrigerant vapor sensors.
    b. contain fire suppression sensors.
    c. have exhaust vent fans.
    d. have both (a) and (c).

63. Purge units should always,
    a. be tested with air.
    b. be vented to mechanical room.
    c. be tested for leaks.
    d. be vented to the outside air.

64. Before converting equipment to other refrigerants, ensure
    a. that there are no leaks.
    b. compatibility with the new refrigerant.
    c. that the manufacturer approves.
    d. (a) (b) and (c)

65. Gas leaving the evaporator is
    a. heated. b. chilled. c. superheated. d. subcooled.

66. Refrigerant blends that do not use mineral oils use
    a. benzacol oils.   b. polyalphitic oils.
    c. aldehide resins.   d. alkylbenzenes.

67. Motor windings, gaskets and seals may be dissolved by refrigerants because
    a. they are powerful solvents.
    b. of sump oil miscibility.
    c. of hygroscopic absorption.
    d. of freon burn acids.

68. Eddy current testing is used to test
    a. eddies.   b. condenser tubes.
    c. transfer plates.   d. silicon buildup.

69. The ability of lubricants to mix with refrigerants is called
    a. Miscibility.   b. Drop point.
    c. Ternary match.   d. Azeotropeability.

70. HFC refrigerants may use what oil(s)?
    a. Alkylbenzenes.   b. Polyalkaline glycols.
    c. Polyolesters.   d. (b) and (c)

71. A component of any refrigerant management plan is
    a. Recovery.   b. Leak testing.
    c. Assess refrigerant needs.   d. Refrigerant tracking.

72. An azeotrope is
    a. A leak testing instrument for HFCs only.
    b. A mixture of 2 or more refrigerants acting as one.
    c. A leak testing insrrumentfor HFCs & HCFCs only.
    d. A leak testing instrument used wih nitrogen.

73. A leak-free system had its CFC-12 replaced with HFC-134A and now leaks. Why?
    a. HFCs are lighter & easily leak when CFCs do not.
    b. HFCs need rubber gaskets.
    c. HFC-134A is corrosive and ate thru the gaskets.
    d. Gaskets were breached by pulling excessive vacuum.

74. A refrigerant having more than one temperature vs pressure curve is said to exhibit
    a. Break down.   b. Freon burn.
    c. Temperature glide.   d. Non-condensables.

75. What lubricant will clean pipe scale, sludge, and other contaminates from appliances?
    a. Mineral oil.   b. Benzacol oil.
    c. Polyolester oil.   d. Aldehide oil.

76. The mixing of three different refrigerants that exhibit separate characteristics under load, though considered one refrigerant, is a (an)
    a. Thermal blend.   b. Azeotrope.
    c. Ternary blend.   d. (b) and (c)

77. Synthetic oils' exposure to air must be minimized because these oils are
    a. Miscible.   b. Aldehides.
    c. Hygroscopic.   d. Water based.

78. The part of an appliance that produces high pressure liquid at low temperatures is the
    a. Compressor.   b. Evaporator.
    c. Condenser.   d. Accumulator.

79. Running a compressor with a deep vacuum will cause
    a. Sticking valves.   b. Burn out.
    c. Brown out.   d. Sublimation.

80. Before recharging an appliance with substitute refrigerants or synthetic oils you should.
    a. Replace compressor   b. Replace accumulator.
    c. Put on SCBA.   d. Replace filters & driers

CHAPTER SEVEN

SECTION 608
GROUP II PRACTICE QUESTIONS

The practice questions are presented in several formats. Similar questions may be found in both multiple choice and true or false formats to help you retain the information. Select the most correct answer for each question. The sample question, number (1A), is an example of a multiple choice format. The example question's correct answer is d. All practice question answers are located in Appendix 1.

Some questions have a reference identified in parentheses directly below the question. If you can't answer the question, go to the reference before looking in Appendix 1 for the answer. If a reference isn't specified, look through the Index and Table of Contents to locate the subject. First try to answer all of the questions in a section before referring to Appendix I.

EXAMPLE QUESTION (1A)

Which of the following statements about the earth is correct?

    a.   The earth is round and hollow
    b.   The earth doesn't rotate around the sun
    c.   The earth is the smallest planet in our solar system
    d.   The earth is round and rotates around the sun

(Our Solar System, Sentinel Publications, 1993)

TEST TAKING STRATEGIES

Review the test taking strategies in Chapter Seven before answering these questions. Use these strategies on the practice tests in this book, on the closed book practice exam in Appendix III, and when you take your actual EPA certification exam.

Here is a (VERY IMPORTANT TIP) for taking the certification tests. The section of the exam you are taking should direct you to the answer. A Type I exam will only have correct answers for a Type I exam, that is, the answers will be for small appliances. Here is an example:

What are the characteristics of a type I appliance?
a. Hermetically sealed, 5 pounds max of charge, drinking water cooler
b. Heat pump, 5 pounds max of charge, open compressor
c. Freezer, 6 pounds of charge, hermetically sealed
d. 5 pounds max of charge, dehumidifier, open compressor

The definition in 82-152,(v) identifies (a) as the correct answer. Answers b,c and d have parts of Type II equipment, open compressors and 6 pounds of charge.

TYPE I PRACTICE QUESTIONS

1. What is the minimum recovery vessel capacity in pounds for Type I equipment?
    a. 6 lbs capacity   b. 5 lbs capacity
    c. 8 lbs capacity   d. 10 lbs capacity

2. This recovery equipment type relies on the appliance compressor for recovery.
    a. Reliant recovery   b. System dependent
    c. Appliance dependent   d. Recovery dependent

3. The best way to service an appliance manufactured after January 1, 1994 is?
    a. Line tap valve   b. Schrader valve
    c. Process stub   d. Pitiot tube

4. What refrigerants may a Type I technician purchase?
    a. All refrigerants
    b. Only those allowed for Type I appliances
    c. Medium pressure only
    d. None of these

5. Why are recovery standards for repair and disposal different?
    a. Repair requires opening the system, disposal does not.
    b. Disposal does not require removal of oil.
    c. There is no difference.
    d. The difference is only due to the manufacturing date of the recovery equipment.

6. EPA requires you to have at least
    a. one piece of recovery equipment.
    b. one piece of certified recovery equipment.
    c. one piece of self contained recovery equipment.
    d. some recovery equipment.

7. What is the smallest size container of R-22 that you may purchase by law ?
    a. Any size available   b. Anything over 5 lbs
    c. 20 lbs or more   d. None of these

8. You have recovered 80% of the refrigerant from an operable window unit being disposed of. What should you do next?
   a. Recover another 10%   b. Evacuate to 10 inHg
   c. Put appliance in landfill   d. Evacuate to 15 inHg

9. Replacing the condenser in a small window air conditioner is a
   a. Minor repair
   b. Major repair
   c. Non recovery operation
   d. Requirement of leak repair

10. An appliance has a leak. Which method is the most accurate for locating leaks?
    a. Halide torch   b. Soap bubbles
    c. Gas sniffer   d. Ultrasonic

11. You just repaired an ice maker that has a factory plate indicating either R-12 or R-134A may be used for the refrigerant. You choose R-134A. What if anything do you need to do before charging?
    a. Nothing, just charge the system.
    b. Pressure test the appliance with 150 psi of nitrogen, check with halide torch for leaks.
    c. Flush system three times with synthetic oil, replace filter and drier, then charge.
    d. Pull a 500 micron vacuum for leak test, if no leaks then charge appliance.

12. Anything that could be reasonably expected to release refrigerant from an appliance to the atmosphere is called?
    a. Major maintenance   b. Opening
    c. Minor maintenance   d. De minimis

13. You service appliances using R-12, R-22 and HFC-134A. Equipment certified by Appendix C will only recover
    a. HFC-134A.   b. HFC-134A and R-12.
    c. R-22.   d. R-12.

14. This method of recovery is best for leaving system oil.
    a. Type C   b. Vacuum   c. Vapor   d. Liquid.

15. The use of high and low side connections for recovery is used with?
    a. Elective equipment   b. Type C equipment
    c. Self contained equipment   d. The push/pull method

16. Your recovery unit keeps shutting off from high pressure. What could be the cause?
    a. Collapsed hose on the high side
    b. Collapsed hose on the low side
    c. Clogged filter
    d. Non-condensables

17. You are purchasing a self contained recovery unit for use with multiple refrigerants. Your unit should have
    a. Low loss fittings
    b. Adjustable low and high pressure cutouts
    c. Replaceable filters
    d. Pump out cycle

18. Long hoses are convenient for reaching service ports. What else do they allow?
    a. Refrigerant loss to atmosphere
    b. Do not have to move heavy recovery equipment as often
    c. Easier to hang up service gauges
    d. Longer recovery time

19. When transporting a disposable service jug to a job, it must
    a. be in a cardboard shipping container
    b. Be secured in the vehicle
    c. Be protected from the sun
    d. (a) and (b)

20. When does recovery take longer?
    a. Summer   b. Winter

21. When may you mix refrigerants?
    a. Never
    b. When topping off a ternary blend
    c. When topping off an azeotrope
    d. When recovering burned refrigerants

22. Which is easier to recover, HCFC-22 or CFC-11?
    a. Neither one, they are both nearly equal in vapor pressure.
    b. HCFC-22 is, because CFC-11 is a low pressure refrigerant.
    c. CFC-11 is, because of hydrogen in HCFC-22 and its ODP number.
    d. HCFC-22 is not, because all CFC refrigerants are high pressure.

23. You are hired by a shop and receive training on their recovery unit using a window air conditioner, when you notice the recovery unit's vessel is white and has a 30 lb capacity. You should.
    a. Immediately stop and replace with a green vessel as you are recovering 5 lbs of R-22.
    b. Continue with the recovery, nothing is wrong.
    c. Replace it with a yellow vessel that indicates used refrigerant.
    d. Replace it with a vessel that has a grey body and a yellow top.

24. Recovery is completed. Where may refrigerant be trapped?
    a. In the appliance accumulation
    b. In the appliance drier
    c. In the recovery hoses
    d. None of these

25. What gas is used in a Halide torch for leak testing?
    a. Halide   b. Bromine   c. Methane   d. Propane

26. What does heating the compressor sump during vapor recovery help?
    a. Helps prelubricate the compressor bearings
    b. Prevents a cold start
    c. Helps lower the vapor pressure
    d. Helps release refrigerant from the sump oil

27. A micron vacuum gauge is very useful for Type II and III technicians, but
    a. it is just as useful for Type I technicians
    b. it is unnecessary for Type I technicians
    c. it can not be used with small appliances
    d. it is only accurate for vacuums up to 15 inHg

28. When do you check your recovery equipment for leaks?
    a. Before each recovery
    b. At least once a day
    c. Once a month
    d. After changing the filters

29. When would you evacuate your recovery machine with another recovery machine?
    a. To recover virgin refrigerant
    b. Before recovering a different refrigerant
    c. Before replacing filters
    d. There is no need for it

30. For these figures, 90/80/4 inHg, which is correct?
    a. Compressor operating/operating with leaks/non-condensables.
    b. Compressor inop/operating with leaks/minimum vacuum.
    c. Compressor with leaks/inoperative/ non-condensables/minimum vacuum
    d. Compressor operating/inoperative/minimum vacuum.

31. Three months ago you charged a heat pump. Now you are back and just added 1.5 lbs of refrigerant to the system and found a slow leak in the condenser. What should you do?
    a. Advise owner of leak, and that EPA requires it to be fixed.
    b. Advise owner of leak and what may eventually happen to the appliance if not repaired.

c. Advise owner of leak and set up schedule for future visits.
    d. Advise owner of leak and that you can not service the appliance again unless the leak is repaired.

31. You are leak testing a new condenser. What refrigerant should be used as a trace charge?
    a. CFC-12
    b. CFC-12 or HCFC-22
    c. HFC-134A & nitrogen
    d. Several ounces of HCFC-22

32. A small leak will usually show up first
    a. on the high side gauge.
    b. on the vacuum pump gauge.
    c. on the low side manifold gauge.
    d. gauges rarely show small leaks.

33. When first charged, the compressor should be
    a. on.   b. off.   c. bypassed.   d. heated.

34. One way to leak test an empty appliance is to?
    a. 150 lbs of dry nitrogen and soap bubbles
    b. 150 lbs of dry nitrogen and halide torch
    c. 150 lbs of HCFC-22 and a sniffer
    d. None of these.

35. You have repaired a heat pump and are ready to recharge it. You should
    a. ask the owner if they want you to reuse the old refrigerant.
    b. recharge the system with virgin refrigerant.
    c. use the old refrigerant and top off with new refrigerant.
    d. use extra virgin refrigerant in the old appliance.

36. You have an empty disposable from liquid charging an appliance. What should you do with the disposable?
    a. Save it for use as an air tank
    b. Use it for recovery of burned refrigerant
    c. Recover any remaining refrigerant, then punch a hole in the vessel
    d. Open the valve and put it in a dumpster

37. One indicator of refrigerant overcharging is?
    a. Excessive compressor current
    b. Super cold evaporator air
    c. Increased superheat
    d. Higher than calculated condenser temperature

38. Hermetically sealed appliances may burn up if,
    a. energized with refrigerant.
    b. energized with a vacuum.
    c. energized at all.
    d. energized with heaters on.

39. As a community service you prepare household appliances for disposal. One thing you must do is
    a. Record who gave you the appliance
    b. Make a sworn statement to the disposer
    c. Recover 90/80/14 inHg
    d. (a), (b) and (c)

40. A calibrated charging cylinder is connected to
    a. The center hose.
    b. The recovery unit's high side.
    c. The high side hose for liquid charging.
    d. The low side hose for vapor charging.

41. Liquid charging requires
    a. Liquid into the suction side
    b. Liquid into the accumulator
    c. The compressor to be off
    d. A calibrated cylinder

42. At 70F R-22 has a pressure of 121.5 psig. What is the psia reading?
    a. 136.2 psia.   b. 106.8 psia.
    c. 121.5 psia.   d. 137.6 psia.

43. A Class I refrigerant is/are
    a. HFC.   b. R-11, R-12, R-123.
    c. R-502, R-123, R-11. d. R-11, R-12, R-502.

44. What refrigerant(s) are compatible with polyolester oils?
    a. HCFC blends.   b. azeotropic CFC blends.
    c. CFCs.   d. HFC.

45. To extract, clean and reuse CFCs without meeting ARI Standard 700-1988
    a. Recover   b. Recycle
    c. Reclaim   d. Reconstitute

46. What is the discharge or disassembly for deposit, dumping, placing, discarding or discharging on land or water?

    a. Pollution
    b. Recovery
    c. Recycle
    d. Disposal

47. You hired 30 EPA certified technicians and rented 35 self contained recovery units for a really big job. What else do you need to do?
    a. Train the technicians on the units and register the units with the EPA
    b. Register the units with EPA
    c. Just start the job
    d. Register the technicians with your wholesale house

48. An EPA certified technician must follow recommended practices, use recovery equipment and carry the EPA certification card when on the job. Also
    a. Post a copy at the place of business
    b. Post a copy at the remote job site
    c. No requirement for posting
    d. Both (a) and (b)

49. What must be on your shipping invoice when sending refrigerants to a reclaimer?
    a. Place of business & emergency phone number
    b. Type of refrigerant and date shipped
    c. Date shipped, phone number and refrigerant
    d. Refrigerant type and quantity

50. How long must you keep records required by EPA?
    a. 1 year b. 3 years c. 5 years d. 7 years

51. What must you keep records on?
    a. Refrigerants purchased, sold, recovered, reclaimed, recycled by quantity, type and service call
    b. Recovery equipment registration and technician training and copy of certification
    c. The number of small cans of R-12 purchased
    d. (a) and (b)

TYPE II QUESTIONS

1. What is the smallest size container of R-12 that you may purchase?
    a. Any size available   b. Anything over 5 lbs
    c. 20 lbs or more   d. None of these

2. What are your recovery equipment requirements?
    a. Either system dependent or self contained
    b. Self contained only
    c. At least one piece of self contained
    d. System dependent only
82-156(b)

3. What refrigerants may you purchase?
    a. Only high and very high pressure refrigerants
    b. All refrigerants
    c. Medium pressure only
    d. None of these.

4. Reliant recovery
    a. is a fast recovery process
    b. may only be used for Type I equipment
    c. does not exist
    d. (a) and (b)

5. Service fittings are required on appliances as of
   a. July 1, 1993
   b. November 14, 1994
   c. November 15, 1993
   d. July 13, 1993.

6. What is the annualized leak rate for retail food and cold storage appliances with 50 or more pounds of refrigerant?
   a. 1.5%   b. 15%   c. 3.5%   d. 35%
   [82-158 (i)]

7. EPA established recovery levels must be met unless
   a. Contamination of the refrigerant would result
   b. There are leaks in the system
   c. There is no "unless," recovery levels must be met
   d. Both (a) and (b)
   82-156(a)(i)(ii)

8. How are recovery requirements decided?
   a. By recovery equipment manufacturing date
   b. By recovery equipment manufacturing date and refrigerant used
   c. By recovery equipment manufacturing date, refrigerant used and quantity of charge
   d. By appliance type
   82-158

9. System dependent equipment may always be used
   a. For R-12 recovery
   b. For R-22 recovery
   c. For any refrigerant recovery
   d. For recovery of 15 lbs or less

10. What level of evacuation is required for MVAC-like appliances?
    a. To or below 102 mm Hg   b. 4 inHg
    c. 15 inHg   d. a and b
    82-156(g)

11. You have certified recovery equipment certified by Appendix B and/or C. Is it legal to use them to evacuate MVAC-like appliances?
    a. Yes, they meet 82-158 requirements
    b. No, they do not meet the requirements of 82-36(a)
    c. Yes, if it is marked as per 82-158(h)
    d. Anything that is certified by EPA may be used

12. Backseated is a term used
    a. to describe service valve position
    b. for a reversed accumulator
    c. for a reversed expansion valve
    d. for a back seat driver

13. The only reason refrigerant boils in the evaporator is
    a. Due to the heat passing over the fins

b. Due to the restrictor or expansion valve
c. Due to suction provided by the compressor
d. Due to the Superheat generated by refrigerant speed

BASIC A/C

14. To remove refrigerant in any condition & storing it is
    a. Recycling   b. Reclaiming
    c. Reprocessing   d. Recovery

15. May you reuse used refrigerant?
    a. Yes if the appliance owner approves
    b. As long as it is used in the same appliance
    c. If used in the same appliance or another appliance owned by the same person
    d. (a) or (b) or (c)

16. Your recovery unit keeps shutting off during a push/pull recovery. What might be the cause?
    a. Non-condensables
    b. Service valve in the port closed position
    c. Clogged recovery unit inlet filter
    d. Any of these

17. For fastest recovery use
    a. A jet pump   b. Short hoses
    c. Large diameter hoses d. Any of these

18. De minimis protection only extends to the technician that follows what?
    a. Required practices and uses recovery equipment
    b. Required practices, uses certified recovery equipment and complies with 82-36
    c. 82-156, 82-154 and 82-158 and/or 82-36
    d. None of these

19. What does the pump out cycle pump out to?
    a. Atmosphere   b. Your trash jug
    c. Another recovery unit d. A recovery vessel

20. You are recovering 100 lbs of refrigerant. You have two 50 lb vessels and two 25 lb vessels. Which ones should you use?
    a. Both 50 lb and one 25 lb b. Both 50 lb vessels
    c. Both 25 lb and one 50 lb d. All four

21. How is a ternary blend charged?
    a. Liquid or vapor   b. Vapor only
    c. Each gas separately   d. Liquid only

22. One mm of Hg equals
    a. 1,000 microns   b. 2,500 microns
    c. 500 microns   d. One mm of Hg

23. After liquid recovery is completed, what is next?
    a. Ship the refrigerant to a recycler
    b. Recover the vapor
    c. Not enough information
    d. Leak test unit then recover vapor

24. When recovering refrigerants from large systems what precautions should you take?
    a. Before recovery determine the capacity of your vessel
    b. Check service valves
    c. Look for areas that refrigerant could be trapped during recovery
    d. All of these

25. When removing and replacing the same refrigerant which recovery unit should be used?
    a. recycling b. reclaiming c. recovery d. (a) & (c)

25. Your recovery unit has a higher low vacuum than required.
    a. That's OK, a stronger vacuum never hurt anything
    b. Your recovery unit could burn up
    c. The deeper vacuum could collapse a capillary tube
    d. The deep vacuum may damage a pressure relief valve.

26. MSDS stands for
    a. Multi Stage Dehydration System.
    b. MonoSulpherDiesteroyl Synthetic
    c. Material Safety Data Sheet
    d. A sulphur dioxide based refrigerant

27. Before recovering a different refrigerant with the same recovery unit you must
    a. Replace the oil   b. Replace the filters
    c. Change the vessel   d. a and c.

28. Push/pull recovery requires
    a. Any self contained recovery unit
    b. A recovery vessel with liquid and vapor valves
    c. The recovery output to a second vessel
    d. All three

29. Your new recovery unit has instructions that conflict with the EPA practices.
    a. Follow the factory instructions and notify EPA
    b. Follow the EPA requirements
    c. Recover using the factory instructions, then repeat using the EPA method
    d. Do not use the equipment

30. R-12 is a
    a. Low pressure refrigerant
    b. Very high pressure refrigerant
    c. High pressure refrigerant

d. Pressure varies by equipment application

31. R-13 is a
    a. Low pressure refrigerant
    b. Very high pressure refrigerant
    c. High pressure refrigerant
    d. Medium pressure refrigerant

32. R-22 is a
    a. Low pressure refrigerant
    b. Very high pressure refrigerant
    c. High pressure refrigerant
    d. Medium pressure refrigerant

33. R-114 is a
    a. Low pressure refrigerant
    b. Very high pressure refrigerant
    c. High pressure refrigerant
    d. Medium pressure refrigerant

Chapter 4

34. The key to MVAC-like is?
    a. Non-road use
    b. Non-road use, mechanical vapor compression, open drive
    c. Agricultural use
    d. Uses R-12 for cooling

35. Retail food & cold storage warehouses are examples of
    a. Transport refrigeration
    b. Process refrigeration
    c. Industrial process refrigeration
    d. Chillers

Chapter 4

36. An appliance with 150 lbs of HCFC-22 must be evacuated to what level using recovery equipment manufactured December 12, 1993.?
    a. 15 inHg   b. 4 inHg   c. 0 inHg   d. 25 inHg

37. An appliance containing 350 lbs of R-502 refrigerant has a leak. To what level must it be evacuated using equipment manufactured before November 14, 1993?
    a. 4 inHg for non leaking parts, 0 psig for leaking
    b. 4 inHg leaking or not
    c. 0 psig for leaking
    d. 10 inHg

38. You added 105 lbs of refrigerant to a cold storage warehouse appliance that has a capacity of 290 lbs. What is the leak rate?
    a. 33% b. 15% c. 30%   d. Over 35%

39. What must you do if the leak rate exceeds the EPA allowance?
    a. Notify EPA   b. Shut down the equipment

c. Evacuate the system   d. Notify the owner

40. Purposely ignoring information about leaks is
    a. Shielding   b. Prudent
    c. A violation   d. Not wise

41. A refrigerated railroad car needed 30 lbs last month by service records and took another 75 lbs this

month. System capacity is 650 lbs. What is the leak rate?
a. 33% b. 15%   c. 30%   d. Over 15%

42. How soon must leaks be repaired?
    a. Within 30 days of notice
    b. Before EPA finds out
    c. Never if an abatement plan is followed
    d. Immediately

TYPE III QUESTIONS

1. Never exceed what pressure level on low pressure equipment?
   a. 0 psia   b. 10 psia
   c. 10 psig   d. 0 psig

2. Why must that pressure not be exceeded?
   a. Purge drum will rupture
   b. Purge compressor will lock rotor
   c. Rupture disk will open, venting refrigerant
   d. Equipment will shut down

3. Which is a low pressure refrigerant?
   a. HCFC-123   b. CFC-12
   c. HCFC-22   d. HFC-134A

4. Evacuation is required to what level with pre November 14, 1993 equipment?
   a. 25 inHg   b. 27,000 microns

c. 25 mmHg absolute   d. 29 inHg gauge

5. A major repair would be
   a. Replacing the accumulator
   b Replacing the expansion valve
   c. Replacing the evaporator
   d. Replacing refrigerant

6. To find a leak between the refrigerant lines and water tubes check,
   a. at the drain plug opening with an electronic tester.
   b. at the purge line output.
   c. with a 150 psi nitrogen charge
   d. the tubes with cork plugs.

7. What refrigerant has a B1 ASHRAE rating?
   a. HCFC-22   b. HCFC-123
   c. HFC-134A   d. R-500

CHAPTER EIGHT

SECTION 609
(MVAC) PRACTICE QUESTIONS

The practice questions are presented in several formats. Similar questions may be found in both multiple choice and true or false formats to help you retain the information. Select the most correct answer for each question. The sample question, number (1A), is an example of a multiple choice format. The example question's correct answer is d. All practice question answers are located in Appendix 1.

Some questions have a reference identified in parentheses directly below the question. If you can't answer the question, go to the reference before looking in Appendix 1 for the answer. If a reference isn't specified, look through the Index and Table of Contents to locate the subject. First try to answer all of the questions in a section before referring to Appendix I.

EXAMPLE QUESTION (1A)

Which of the following statements about the earth is correct?

    a.   The earth is round and hollow
    b.   The earth doesn't rotate around the sun
    c.   The earth is the smallest planet in our solar system
    d.   The earth is round and rotates around the sun

(Our Solar System, Sentinel Publications, 1993)

TEST TAKING STRATEGIES

Review the test taking strategies in Chapter Seven before answering these questions. Use these strategies on the practice tests in this book, on the closed book practice exam in Appendix III, and when you take your actual EPA certification exam.

TEST QUESTIONS (SECTION 609)

NOTE: Also review the Group I or core questions in Chapter Six to learn about ozone, ultraviolet UV-B rays, etc.

1. The only time it is permissible to mix R-12 and R-134A is
   a. When converting the MVAC to an R-134A charged system.
   b. When evacuating the system.
   c. When recovering burned or acid contaminated refrigerant for disposal.
   d. It is never permissible.

2. Which is the proper way to dispose of an empty disposable?
   a. Evacuate with a recovery/recycling unit until it shuts off, close the valve and put in dumpster.
   b. Recovery/recycling evacuation, close valve, punch a hole in cylinder.
   c. Vacuum pump to at least 26inHg, put in trash.
   d. Put in dumpster or use as portable air tank.

3. Refrigerated cargo vehicles with hermetically sealed systems are
   a. Covered only if containing CFC-12.
   b. Not covered if using CFC-12.
   c. Not covered.
   d. Covered if containing CFC-12 and HCFC-22.

4. R-12 may be reused, provided that it is cleaned to specifications in?
   a. SAEj1991.
   b. SAEj1990.
   c. SAEj1989
   d. SAEj1992.

5. During operation liquid fill of recovery tank shall not
   a. Exceed 60% of tank capacity at 75°F.
   b. Exceed 80% of tank capacity at 70°F.
   c. Exceed 60% of tank capacity at 70°F.
   d. Exceed 80% of tank capacity at 75°F.

6. Certified extraction equipment must evacuate to what level?
   a. 26in Hg.   b. 27in Hg.
   c. 28in Hg.   d. 29in Hg.

7. A standard that all portable refillable tanks must meet
   a. SAEj 1991.
   b. UcLa.
   c. DOT.
   d. 29CFR 1910.

8. Non-condensables cause
   a. Acids.
   b. High head pressure.
   c. Oxidation of oil.
   d. Purging.

9. Equipment that handles R-12 must
   a. Be EPA approved.
   b. Comply with applicable Federal, State and local requirements.
   c. Require training.
   d. Not purge.

10. Your recycling equipment indicates the moisture content is too high for the R-12 you are processing. What must you do?
    a. The purge cycle will take care of it, do nothing.
    b. The refrigerant is too contaminated, replace it.
    c. The refrigerant may be too hot for proper sensing, wait 20 minutes for system to cool then retry.
    d. Replace the units filter/dryer.

11. Equipment certified before February 1, 1992 must meet?
    a. SAEj 1991.   b. SAEj 1990 (1989).
    c. SAEj 1992.   d. SAEj 1991 (1990).

12. Refrigerant directly removed from and returned to a MVAC is
    a. Recovered
    b. Processed
    c. Recycled
    d. Not allowed.

13. When must tanks be retested?
    a. Every 4 years.
    b. Every 5 years.
    c. After each 400 uses
    d. Only if you suspect a leak

14. Aside from refrigerant release why are shutoff devices required?
    a. To reduce non-condensables.
    b. To prevent reuse of refrigerant.
    c. To reduce non-condensables into the recovery equipment.
    d. To allow quick shutoff if refrigerant is contaminated.

15. When should used lubricant be recharged into the system?
    a. When liquid charging.
    b. Never.
    c. Separately from the liquid charging.
    d. After filtering.

16. After recovery from a system you have more oil than you charged. You had flushed the system, replaced the compressor, accumulator and drier and personally recharged this system with measured refrigerant and oil. Where did the extra oil come from?
    a. Refrigerant dissolved in the oil.
    b. Oil in the compressor.
    c. Shipping oil in the accumulator.
    d. Oil trapped in the evaporator.

17. What document provides equipment specifications for CFC-12 recycling equipment manufactured after February 1, 1992 ?
    a. SAEj 1992.
    b. SAEj 1990(1990).
    c. SAEj 1991(1990).
    d. SAEj 1990(1991)

18. Entities which service MVACs for consideration must
    a. Keep all records for 3 years.
    b. Keep all records on site.
    c. Keep records.
    d. (a) and (b).

19. An alternative for R-12 is
    a. HFCF-22.   b. HCFH-22.
    c. HFC-134A.   d. HFCF-134A.

20. The standard for service hoses is
    a. SAEj 1990.
    b. SAEj 2196.
    c. SAEj 1991.
    d. FF-12.

21. Hoses for R-12 and R-134A are
    a. Interchangeable.
    b. Non-interchangeable.
    c. Neoprene rubber.
    d. Braded tygon rubber.

22. Polyolester oil may be used with
    a. R-12.
    b. CFC refrigerants.
    c. HFC refrigerants.
    d. HCFC refrigerants.

23. December 31, 1995 is when?
    a. All refrigerants must be non chlorinated.
    b. All refrigerants must be Hydrogenated.
    c. All CFC refrigerant production stops.
    d. HCFC refrigerants must be used.

24. The date that technicians are required to be certified to work on MVACs.
    a. July 1, 1992.
    b. August 13, 1992.
    c. June 29, 1992.
    d. January 1, 1993.

25. Can Substituting R-134A for CFC-12 be done without system modification?
    a. Yes   b. No

26. For recovery if there is pressure after_____, then repeat the operation until the vacuum remains stable for _____.
    a. 5 min, 5 min.
    b. 3 min, 2 min.
    c. 5 min, 1 min.
    d. 5 min, 2 min.

27. When are manifold hose service valves closed?
    a. Anytime not in use.
    b. Before disconnecting from A/C system.
    c. During purge operation.
    d. None of the above.

28. Testing for non-condensable gasses requires a
    a. Set of scales.
    b. A temperature above 65°F.
    c. Manifold gauge.
    d. Container temperature reading.

29. Where are non-condensables vented to?
    a. Atmosphere because they are moisture and air.
    b. A disposable tank.
    c. In to recycling equipment.
    d. Cannot be vented, must be recycled again.

30. Prior to use, external portable containers must be
    a. Evacuated to 26in Hg.
    b. Evacuated to 27in Hg.
    c. Evacuated to 28in Hg.
    d. Evacuated to 28in Hg.

31. Any container of recycled refrigerant that has been stored or transferred must be tested for what?
    a. Non-condensables.
    b. Cylinder pressure.
    c. Leaks.
    d. Weight of contents.

32. MVACs using HFC-134A have
    a. Smaller evaporators.
    b. No oil.
    c. Quick connect connectors
    d. Threaded connectors.

33. The standard SAEj 1989 is for
    a. HFC-134A.   b. HCFC-134A.
    c. CFC-12.   d. HFC-12.

34. The standard SAEj 2211 is for
    a. A fake number. b. HFC-134A.
    c. CFC-12.   d. HCFC-134A.

35. What is the first thing to do before recovery?
    a. Make sure the system has a charge.
    b. Make sure the customer can pay.
    c. Make sure the system is cool.
    d. Chill your recovery vessel.

36. What must you do after recovering R-12 and before recovering R-134A?
    a. Change the oil.
    b. Change the oil, clean the unit.
    c. Clean the unit, change the oil, replace the filters.
    d. Change the oil, replace the filters, clean the unit, change hoses.

CHAPTER NINE

RECOVERY RECYCLING
PRACTICE QUESTIONS

The practice questions are presented in several formats. Similar questions may be found in both multiple choice and true or false formats to help you retain the information. Select the most correct answer for each question. The sample question, number (1A), is an example of a multiple choice format. The example question's correct answer is d. All practice question answers are located in Appendix 1.

Some questions have a reference identified in parentheses directly below the question. If you can't answer the question, go to the reference before looking in Appendix 1 for the answer. If a reference isn't specified, look through the Index and Table of Contents to locate the subject. First try to answer all of the questions in a section before referring to Appendix I.

EXAMPLE QUESTION (1A)

Which of the following statements about the earth is correct?

    a.   The earth is round and hollow
    b.   The earth doesn't rotate around the sun
    c.   The earth is the smallest planet in our solar system
    d.   The earth is round and rotates around the sun

(Our Solar System, Sentinel Publications, 1993)

TEST TAKING STRATEGIES

Review the test taking strategies in Chapter Seven before answering these practice questions. Use these strategies on the practice tests in this book, on the closed book practice exam in Appendix III, and when you take your actual EPA certification exam.

RECOVERY / RECYCLING
PRACTICE QUESTIONS

1. Holding cylinders and storage tanks should not be liquid filled past what capacity?
    a. 90% of volume @70°F
    b. 80% of volume @70°F
    c. 75% of volume @70°F
    d. 95% of volume @70°F
Section 609, 82, Appendix A, 7.3, 7.4

2. Refrigerant cylinders, jugs or tanks may burst when heated to 130°F due to
    a. gas pressure.
    b. air pressure.
    c. hydrostatic pressure.
    d. thermo pressure.

3. Refrigerant cylinders, jugs or tanks may burst at room temperature if
    a. charged to 100% of capacity.
    b. pressure actuated relief valves are installed.
    c. subjected to vibration.
    d. stored on their sides.

4. Leaks will usually show up where?
    a. On the high side pressure gauge
    b. On the low side pressure gauge
    c. In the sight glass
    d. As a - 14.7 psig reading

5. A name given to refrigerants removed from systems for recovery, recycling, storage, reprocessing or transportation is
    a. recovered waste.
    b. recovered refrigerant.
    c. purged refrigerant.
    d. standard sample.
608, 82, Subpart-F, Appendix B, 3.1

6. A specified requirement, that is mandatory for compliance with a standard will contain the word _____, in the statement.
    a. maybe   b. should
    c. always   d. shall
608, 82, subpart-F, appendix B,3.9.1

7. The presence of Non-condensable gasses in storage tanks may indicate poor quality control in
    a. manufacturing.
    b. cleaning tanks.
    c. recovery equipment.
    d. transferring refrigerants.
608, 82, subpart-F, appendix A, 5.9.1

8. Appendix A to subpart-F, specifications for fluorocarbon refrigerants is based on Air conditioning and Refrigeration Institute Standard,
    a. BB-F-1421.
    b. 700-1988.
    c. J-1990.
    d. 740-1991.
608, 82, Appendix A

9. A thermistor vacuum gauge is used to measure vacuum in
    a. absolute pressure.   b. gauge vacuum.
    c. microns.   d. none of these.

10. What percent of refrigerants must be recovered when opening a small appliance for repair? The compressor is inoperative and your recovery equipment was manufactured January 1, 1994.
    a. 80%   b.85%
    c.90%   d.95%
82.156,(a)(4)(ii)

11. What is the effective date of the law that requires you to have and use recovery equipment and that you are complying with the rules?
    a. June 14, 1993
    b. November 15, 1993
    c. July 13, 1993
    d. August 12, 1993
82.154(e)(f)

12. Hydrostatic means
    a. Incompressible fluids   b. Water energy
    c. Frozen pipes   d. Moving liquid

13. The standard used to evaluate ALL refrigerants for section 608 is
    a. JFP-621.   b. j-2 titration.
    c. ARI Standard 700-1988.   d. j-621.
82.164(a)(b), section 608, Appendix A to subpart F

14. A word used to indicate a procedure which is desirable as good practice but not mandatory.
    a. Maybe
    b. Sometimes
    c. Shall
    d. Recommended
Appendix A to Subpart F of 608

15. The most accurate vacuum reading is taken with a
    a. manifold gauge set.
    b. mercury barometer.
    c. an absolute gauge.
    d. micron gauge.

16. A liquid phase sample is NOT used in testing for
    a. acidity
    b. water
    c. non-condensables
    d. chlorides
Appendix A to Subpart F,5.2.3

17. Non-condensable gasses are usually made up of
    a. air.
    b. trace refrigerants.
    c. chloride gasses.
    d. oil vapors.
Appendix A to subpart F, 5.9.1

18. Compliance to ARI Standard 700-1988 by manufacturers, reclaimers and repackagers of fluorocarbon refrigerants is
    a. mandatory.
    b. voluntary.
    c. conformal.
    d. heuristic.
Appendix A to subpart F, 7.1

19. Refrigerant that has been removed from a system for the purpose of storage or transportation is
    a. recycled refrigerant.
    b. reclaimed refrigerant.
    c. recovered refrigerant.
    d. removed refrigerant.
608, Appendix B to subpart F,3.1

20. Recovery equipment which requires the assistance of components contained in an air conditioning or refrigeration system is called
    a. self contained.
    b. general equipment.
    c. push/pull equipment.
    d. system dependent equipment.
608, Appendix B to subpart F, 3.8.2

21. A recovery method which lowers the pressure in the vessel and raises the pressure in the system is the
    a. push/pull method.
    b. vacuum/push method.
    c. incorrect method.
    d. self contained method.
608, Appendix B to subpart F, 3.6

22. The low pressure gauge of a manifold gauge set
    a. is marked red.
    b. is marked red and labeled LOW.
    c. is marked blue.
    d. is which ever one you select as they both measure vacuum accurately.

23. Refrigerant recovered divided by refrigerant recoverable times 100% is termed,
    a. refrigerant recovered.
    b. recovery final.
    c. refrigerant recoverable.
    d. recovery efficiency.
Appendix C to subpart F

24. Refrigerant recovered equals, (pick the MOST correct)
    a. SCF-SCO+RSF-RSO
    b. SCF+SCO-RSF+RSO-OL
    c. SCF+SCO+RSF-RSO
    d. SCF-SCO-RSF-RSO
Appendix C to subpart F,IV Calculations

25. The amount of refrigerant processed (in pounds) divided by the time elapsed in the recycling mode in pounds per minute is the
    a. push/pull rate.
    b. pressure filter rate.
    c. speed ratio.
    d. recycle rate.
Appendix B to subpart F, 3.7

26. This term usually implies the use of processes or procedures available only at a reprocessing or manufacturing facility.
    a. Recycle
    b. Reclaim
    c. Recovery
    d. Reprocess

Appendix B to Subpart F, 3.4

27. You have a registered, system dependent recovery system and you are about to open an appliance that contains 22 lbs of refrigerant. You should,
    a. evacuate the system and make necessary repairs.
    b. obtain self contained equipment to evacuate the system.
    c. use both equipments push/pull style for speed.
    d. recover the refrigerant to a 25 lbs service jug.

82.156,(b)(c)

28. To recover 22 lbs of refrigerant what would be the minimum capacity of your recovery vessel?
    a. 22 lbs.   b. 30 lbs.
    c. 25 lbs.   d. 23 lbs.

29. Type III technicians would use this method of recovery:
    a. inflatable bag.   b. push/pull.
    c. system dependent.   d. inverse purging.

30. A recovery units high pressure cut out switch helps prevent
    a. sublimation of refrigerant.
    b. acids from freon burn.
    c. damage to recovery unit.
    d. increased recovery time.

CH5

31. Liquid charging is done
    a. from the charging cylinder through the center manifold and out the high side manifold to the suction side of the appliance.
    b. from the charging cylinder through the center manifold and out the high side gauge to the appliance liquid line.
    c. from the charging cylinder through the center manifold to the low side manifold to the suction side.
    d. from the charging cylinder direct to the liquid line.

32. Defrost heaters will help with
    a. Preventing freon burn
    b. System dependent recovery

    c. Refrigerant release from oil
    d. (b) and (c)

33. Before using any recovery vessel, what should be done?
    a. Evacuate it of all air and moisture, check for UL or DOT approval.
    b. Evacuate it of all air and moisture, check for UL or DOT approval, establish maximum safe capacity of vessel.
    c. Evacuate it of all air and moisture, check for DOT approval, check for signs of freon burn.
    d. Evacuate it of all non-condensables and moisture, check for UL approval.

34. The most accurate method for charging is to use
    a. Balance beam scales
    b. A manifold gauge set
    c. Electronic scales
    d. Temperature chart

35. Apping is a term for
    a. Releasing refrigerant from oil.
    b. A form of musical expression.
    c. Heat pump knock.
    d. Recovery unit failure.

36. As of this date you must have EPA certified recovery equipment.
    a. July 13, 1993
    b. November 14, 1994
    c. July 1, 1993
    d. November 15, 1993

37. You have isolated a compressor for evacuation and removal. Factory manual indicates a charge of 35 pounds of HCFC-22 when isolated. What is the required level of evacuation using equipment manufactured after November 15, 1993?
    a. 15 inHg
    b. 4 inHg
    c. 25 mm Hg abs
    d. 0 inHg

38. You have started a recovery and have noticed that the gauge is not dropping. What does it indicate, if anything?
    a. A very large charge.
    b. Non-condensables.
    c. A big leak.
    d. Open the gauge chamber.

39. A major repair would be removal of
    a. Evaporator, compressor, condenser and/or auxiliary heat exchanger coil.
    b. Evaporator, compressor, accumulator, condenser and/or auxiliary heat exchanger coil.
    c. Evaporator, compressor, restrictor and condenser.
    d. Evaporator, condenser, auxiliary heat exchanger and accumulator.

40. The center manifold chamber on service gauges may be connected to
    a. the high side, the recovery unit and the disposable service jug.
    b. the low side, the recovery unit and the disposable service jug.
    c. the high side, the recovery unit and the charging cylinder.
    d. the recovery unit, the vacuum pump and the charging cylinder.

41. You discover a leak on a heat pump that has a 9 lb charge of R-22. Calculations show a leak rate of 22% per year. How soon must it be repaired?
    a. No requirement to repair
    b. Within 30 days of your service call
    c. 30 days or 1 year if abatement plan is developed
    d. NO requirement as rate does not exceed 35%

42. To recover 300 lbs of R-502 from a non-leaking appliance with equipment manufactured before November 15, 1993, what evacuation level must be met?
    a. 4 inHg
    b. 4 inHg or 0 psig
    c. 15 inHg
    d. 15 inHg or 0 psig

43. A leaking appliance using HCFC-123 is to be evacuated to what level?
    a. 0 inHg
    b. 10 inHg
    c. Pressurized to 0 psig
    d. 4 inHg

44. Water has obscured labeling tags on two recovery vessels. Using established practice, tank A has a temperature of 78°F and a pressure reading of 139 psig. What refrigerant is in tank A?
    a. R-134A
    b. R-502
    c. R-500
    d. R-22

45. A small working appliance containing 3.5 lbs of CFC-12 is being disposed of. How much refrigerant must be recovered?
    a. 2.9 lbs
    b. 50.4 oz
    c. 2.7 lbs
    d. None if there aren't any leaks

46. You have just repaired an MVAC-like appliance. The owner requests you "top off" his automotive air conditioner while you are there. You
    a. can do it as you are certified as a Type II technician.
    b. must be certified as a Type Universal technician to do the work.
    c. must be certified under section 609 to do the work.
    d. are covered by the 608 blanket waver for MVACs servicing.

47. A two ton window unit requires evacuation. What method is least likely to remove system oil?
    a. Push/pull
    b. Vapor
    c. Liquid
    d. dehydration

48. Sublimation is caused by
    a. freon burn residue
    b. too large a vacuum pump
    c. acid flux
    d. (a) and (c)

49. You are using a 7CFM recovery unit to recover vapor from an isolated chiller unit. What precautions should you take, if any?
    a. The rate of evacuation may cause the chiller water to freeze and rupture the pipes.
    b. Evacuate away as the vapor pressure is high enough that freezing cannot happen.
    c. You must use a flow restrictor valve to prevent possible freon burn.
    d. Chiller tubes may collapse due to the extreme vacuum generated by any recovery equipment above 5CFM capacity.

50. What vacuum level should be obtained to ensure non-condensables are removed?
    a. 0 psig
    b. 500 microns
    c. 30 in Hg
    d. 30 microns

51. The recovery machine is connected to the
    a. Center chamber of the gauge set.
    b. Right or low side chamber.
    c. Left or high side chamber.
    d. Right or blue gauge side.

52. Before evacuating an appliance
    a. bring system pressures up using nitrogen.
    b. leak test all connections.
    c. check sight glass for bubbles.
    d. replace the freon burn indicator on the inlet line.

53. After the vapor recovery unit shuts off how long should you wait before disconnecting the recovery equipment?

    a. Several minutes or two cycles of the recovery equipment.
    b. Disconnect immediately.
    c. Five minutes and no more than a 1,000 micron decrease in vacuum.
    d. Five minutes and no more than a 500 micron decrease in vacuum.

54. A recovery vessel may be
    a. a disposable jug.
    b. filled to full rated capacity.
    c. a grey cylinder with a yellow top.
    d. a disposable jug painted yellow and grey.

55. Refrigerant recovers faster when
    a. warm.   b. cold.

RTI MaxiFlex TX 600 Recovery/Recycling/Refining
and MaxiLite Recovery TX200 Systems
Complements of RTI Technologies, Inc. (800-469-2321)

CHAPTER TEN

PREPARING FOR EPA INSPECTIONS

What to do Before, During, and After You Are Inspected

Everyone dreads an **"INSPECTION"**. It doesn't matter what kind it is, when the word **"INSPECTION"** is used people tense up, start worrying and imagine the worst. On the other hand, the inspector never knows exactly how he or she will be treated when they arrive on-site.

Repair shops must expend the resources necessary to comply with the Clean Air Act. You must first know the applicable regulations, laws, and recovery for freon handling methods before you can effectively plan for an inspection. The first step is to thoroughly read and understand the Clean Air Act. This involves training the technicians and preparing them for EPA certification. After the workforce is trained and prepared to service freon based systems under the new rules, a company then must assess its Clean Air Act compliance strengths and weaknesses.

Section 608, 82-154(a) of the Clean Air Act (CAA) states that, **"effective June 14, 1993**, it shall be unlawful for any person, in the course of maintaining, servicing, repairing, or disposing of an appliance or industrial process refrigeration, to knowingly vent or otherwise knowingly release or dispose of any class I or class II substance used as a refrigerant in such appliance (or industrial process refrigeration) in a manner which permits such substance to enter the environment." The EPA is responsible for enforcement of this regulation.

Section 609 of the CAA states that, "Any service involving the *motor vehicle air conditioner refrigerant* requires the refrigerant to be recycled by certified technicians using certified equipment. Certified recycling equipment includes recover/recycle and recovery only equipment. Also, the sale of Class I and Class II substances, CFCs and HCFCs respectively, in containers less than 20 pounds is restricted to certified technicians. These containers are referred to as "small cans". The sales restriction on small cans is effective **November 15, 1992**. There is no restriction on the sale of containers weighing 20 pounds or more." Refer to Section 609's enforcement and compliance guidance beginning on page 121.

It is estimated that there are currently 200,000 establishments performing service on motor vehicle air conditioners. The number of technicians that need to be certified is estimated to be over 500,000.

CAUTION

The interim Clean Air Act penalty and enforcement guidance is provided in this chapter to show how fines are assessed and inspections are handled by the EPA. The information in this book, including the interim guidance, does not replace or supersede official EPA guidance, memorandums, final decisions, and the CFR. It is important to check with EPA offices for the most up-to-date guidance when you are developing your compliance plans and preparing for on-site inspections. The consequences of not having the most current information could lead to fines as high as $25,000 per violation per day.

The author and publisher make no warranties, either expressed or implied, with respect to the information contained herein. The information about inspections and EPA regulations and laws were obtained from various EPA resources including electronic bulletin board computer files, EPA Memorandums, the CFR, interim guidance letters, and from discussions with EPA personnel. The author and publisher shall not be liable for any incidental or consequential damages in connection with, or arising out of, the use of materials in this book.

PREPARING FOR INSPECTIONS

The following checklists were designed to assist shops with inspection preparation. They will also help you assess your shop's EPA compliance strengths and weaknesses. Use the following checklists and the enforcement and compliance sections in this chapter as a starting point. Add, subtract, and modify your compliance plans and checklists until they reflect the true picture of where you and your company stand with the many compliance issues. After your internal evaluation, develop action plans and delegate corrective actions as necessary. The more technicians and managers involved with this process the better it will work. Don't forget to have numerous follow-up meetings to check progress and to revise the plan as situations dictate.

What to do when the inspector arrives

☐   Check the inspector's identification.
(All official inspectors will offer identification upon arriving on-site.)
☐   Introduce yourself and others in the room.
✔ Offer coffee and show where the rest rooms are.
✔ Treat the inspector like you would your customer.
☐   Always remember that everything you say is — **ON THE RECORD.**
☐   Listen and take notes during the entrance briefing.
☐   Never — **NEVER** — argue with the inspector.
☐   Accompany the inspector during the inspection.
☐   Introduce the inspector to everyone in the shop.
☐   Insure that everyone cooperates with the inspector.
☐   Ask the inspector for recently released EPA updates and Fact Sheets.
☐   If you are cited for violations ask for suggested remedies and references.
✔ What is the proposed fine, if any?
✔ What must be done to comply and by when?
☐   Ask the inspector if you can call him /her direct for guidance and request a phone number where he/she can be reached.

DEVELOPING INSPECTION CHECKLISTS

The best defense is a good offense — **BE PREPARED.** Federal, state and city inspectors usually follow checklists. Checklists provide inspectors with a systematic evaluation process to assess compliance as well as to standardize inspections. Your preparation, or lack of it, may well determine whether or not you will successfully pass the inspection and avoid fines.

Fines can be levied against companies and individual employees.

It's important to develop tailored inspection check lists for your company or shop long before an inspector arrives. Inspection lists allow companies to assess their organization for potential weaknesses and at the same time provide a plan-of-attack to resolve any discrepancies. You could list everything in the Code of Federal Regulations part 40CFR.82.XXX or develop a generalized checklist for specific areas.

After reading this book, you will be better prepared to evaluate your inspection readiness. The shop technicians know better than anyone what EPA regulations have been emphasized and which ones have slipped through the cracks. Let's begin by trying not to overlook the obvious.

BRAINSTORMING " *THE OBVIOUS* "

OK Fred, Why do we need to review the obvious?

Humans tend to forget the obvious. Once missed or forgotten, the obvious returns to inflict mental anguish, injury or monetary loss. It's embarrassing to get nailed for the obvious. Prepare effectively and avoid embarrassment by listing every area that is suspect and needs attention.

Use the process called brainstorming to develop your lists and compliance strategies. To brainstorm a project, participants think out loud about how they would accomplish the task. Any idea or word is listed and written down without judgement, no matter how out-of-line the original idea may first seem. Brainstorming can be done by one person or in groups. Once the ideas and words stop flowing it's time to analyze what was listed. Look for ideas that have a common thread and those that could be combined or grouped into similar areas. This process may result in other lists or sub groups. Discuss the lists until you run out of ideas, then develop YOUR CHECKLIST.

A checklist, developed from the brainstorming process, should capture many of the items that an inspector would look for and it will often reveal areas that need attention. The following *Pre-Inspection Check List* outlines many areas that you may wish to investigate prior to EPA inspections. Add items from your brainstorming sessions to your personal check list and assign individuals the responsibility to check out specific areas or develop more detailed lists. Don't forget to have followup meetings to review how things are going. Here are some items to help you get started:

THE "BOUNTY" PROVISION

EPA can pay an award, not to exceed **$10,000**, to any person who furnishes information or services which lead to a criminal conviction or a civil penalty assessed as a result of a violation of the CAA.

Pre-Inspection Checklist
(FILL IN THE BLANKS & EXPAND LISTING)

☐ Violations of prohibitions, i.e. (partial list)
- ✔ Location of shutoff values
- ✔ Registration of recycling equipment
- ✔ CFC purchases & storage
- ✔

☐ Record keeping violations, i.e. (partial list)
- ✔ Certifications displayed in shop
- ✔ Freon purchasing records
- ✔

☐ Following recommended practices, i.e. (partial list)
- ✔ Recovery methods
- ✔ Documentation
- ✔

☐ Use of (3R) recovery, reclaiming, recycling equipment, i.e. (partial list)
- ✔ Is recovery/recycling equipment available & certified
- ✔ Are there any unauthorized modifications to your 3R equipment?
- ✔

☐ Container markings, i.e. (partial list)
- ✔ CFC containers
- ✔ Used oil/fluids

☐ Are copies posted of each technicians certification card?
☐ Are technician carrying their certification card while on the job?
☐ Does the company support Section 608/609 requirements? i.e. (partial list)
- ✔ Disposal of appliances with CFCs
- ✔
- ✔

☐ Are technicians complying with section 608/609 in the shop and with customers?

Pre-Inspection Checklist (Cont.)

☐ Does everyone use:
- ✔ Use safety equipment
- ✔ Follow safe working practices
- ✔ Use proper materials on the job

   NOTE: These areas should be included in your shop training plan.

☐ Is there a shop training plan? (Reference Appendix VIII for sample plan)

☐ Safety hazards; chemical, mechanical, physical, i.e. (partial list)
- ✔ Are Material Data Safety sheets (MSDS) available for chemicals used?
- ✔ Are copies of recent site safety inspections available?
- ✔

☐ First aid and CPR training. Even if emergency medical services are available.

☐ Employee/employer compliance responsibilities, reporting hazards, spills, OSHA and RCRA.

☐ Disposal of used recovery oils and filters, recycling and reprocessing equipments or refrigeration systems.

☐ Who has what responsibilities in the shop?
- ✔ Is there a list of delegated responsibilities posted?

☐ Refrigerants purchase process, i.e. (partial list)
- ✔ Who is purchasing refrigerants? The purchaser must have CFC certification.

☐ Is new equipment registered with EPA?

☐ _____

☐ _____

☐ _____

☐ _____

☐ _____

☐ _____

THE NOT SO OBVIOUS

Now that you've listed many of the obvious inspection areas, you need to think about the areas that will "showcase" your operation. A shop that takes visible preventive steps to ensure compliance now and in the future I believe would be considered a showcase operation. Documentation of the actions below will show inspectors that a positive corporate attitude exists towards EPA compliance. Use the same brainstorming technique as before. Here is a checklist to get you started.

The Showcase Checklist

☐ Document new equipment, training, procedures, and refrigerants.

☐ Review in-use equipment manuals, procedures, and practices.

☐ Send technicians to workshops and factory demonstrations.
  - ✔ If factory demonstrations or workshops aren't scheduled, call the factory and sponsor a workshop. Ask if the company will pick up the cost. If not, will other area shops share the expenses?

☐ Schedule recurring training.

☐ Develop formal training plans for all technicians.
  - ✔ Plans should list each piece of equipment covered by EPA certification requirements and all other essential equipment. **A sample training plan is available in Appendix VII.**

☐ The training plan should include hands-on practical demonstrations.

☐ Post wall charts that show equipment hook-ups.

☐ List all special equipment and procedures that are needed to service and repair your customers' refrigeration and air conditioning systems.
  - ✔ Technicians must check this list prior to departing the shop or radio dispatchers must be able to relay these instructions over the air.
  - ✔ Flag work orders that require special equipment and procedures.

☐ Review NEW in-use equipment manuals, procedures, and practices. ESPECIALLY if procedures are similar to old equipment still being used.

☐ If operating procedures are confusing, attach a laminated card to the equipment with the correct procedures.

☐ If you sponsor a workshop invite an EPA inspector to attend. The inspector can provide updated information about the latest substitute refrigerants, proposed rules and the top ten citations being issued.

SECTION 608/609 SUPPORT
Maintaining a Professional Attitude

" ACTIONS SPEAKS LOUDER THAN WORDS"

The old adage, "action speaks louder than words" applies to most of what we do. Is it enough to have recycling/reclaiming equipment when no one knows how to use it? The following list will help you avoid complacency in your shop. It isn't enough to have certification if you don't follow the rules.

MAINTAINING A PROFESSIONAL ATTITUDE

1. Does management actively support section 608/609?

2. Do you have training that reinforces proper attitudes?
    - ✔ When the shop is training technicians on new systems, are ALL 608/609 recommended practices followed?
    - ✔ Is 3R equipment available and next to the system being serviced, as EPA requires?

3. Do technicians **ALWAYS** follow recommended practices or do you occasionally take short cuts?

4. When servicing refrigeration equipment, is 3R equipment on site, or in the truck? If a leak develops — **whatever the reason** — how would you explain not having it to minimize the venting?

5. Do management decisions support Section 608/609?

6. Do employees put Section 608 ahead of personal convenience?

7. Do employees demonstrate high standards of professional conduct? Does the company?

EPA Inspection Checklists

The official EPA Inspection Lists (Section 608) are provided for your review. There are two Checklists, Level 1, and Level 2. Note block 23 of the Level 1 inspection. It asks if there is a reason to go deeper with a Level 2 inspection and if so list the reasons. The Level 2 inspection is more in depth.

Level 1 Inspection Checklist (Section 608)
National Recycling and Emission Reduction Program

1. Facility Name _____

2. Facility Address _____

   _____

3. Name of Owner/Operator _____

4. Name of Person in Charge _____

5. Title of Person in Charge _____

6. Facility Description

   _____ Food Service/retail food   _____ Service contractor
   _____ Supply house   _____ Reclaimer
   _____ Industrial facility   _____ Landfill/recycler/salvage yard
   _____ Other _____

7. Does your business service, maintain, repair, or dispose of air
   conditioning or refrigeration equipment?   ___ Yes ___ No

8. Types of equipment serviced, maintained, repaired, or disposed of
   (check all that apply).

   ____ Small appliances (household refrigerators, dehumidifiers, vending
        machines, water cooler, other equipment with charge of less than 5 pounds.)

   ____ High-pressure and very high pressure appliances (includes appliances
        using R-12, R-13, R-22, R-114, R-500, R-502, R-503)

   ____ Low-pressure appliances (includes appliances using R-11, R-131, R-123)

   ____ MVAC-like appliances (construction and farm vehicles, other vehicles
        using R-12)

9. Has the facility submitted an EPA Recovery or Recycling Device
   Acquisition Certification?   _____ Yes   _____ No

10. Do they have a file copy?   _____ Yes   _____ No

11. Is Recovery or Recycling Equipment Present?_____ Yes_____ No

12. Type of equipment _____ Recovery/Recycling   _____

13. Is equipment grandfathered? (Manufactured before 11/15/93)
    ___ Yes ___ No

14. Did inspector pull a vacuum with calibrated pressure gauges?
    ___ Yes ___ No

15. Inches of vacuum pulled _____

16. Does equipment meet EPA standards?   ___ Yes ___ No

17. Is equipment laboratory-approved?   ___ Yes   ___ No
    (manufactured on or after November 15, 1993)

18. Equipment brand/Model/Date Manufactured _____

19. Serial Number _____

20. If service is subcontracted, are invoices or purchase orders
    issued to the subcontractor available?   ___ Yes   ___ No

21. Effective 11/14/94: Are technicians certified by an accredited
    program? (check certificates)   ___ Yes   ___ No

22. Type of Certification

I: Number of Technicians ___   Name of Training Course: _____
      Date Attended/Certification Number(s) _____

II: Number of Technicians ___   Name of Training Course: _____
      Date Attended/Certification Number(s) _____

III: Number of Technicians ___   Name of Training Course: _____
      Date Attended/Certification Number(s)_____

Universal: Number of Technicians ___ Name of Training Course:_____
Date Attended/Certification Number(s) _____

Other: Number of Technicians ___ Name of Training Course: _____
       Date Attended/Certification Number(s) _____

23. Does the inspector have reason to expand the inspection?
    ___ Yes (if yes, see Box # on Level 2) ___ No

24. Inspector Name _____ Inspection Date: _____

25. Inspector Signature _____ Inspection Time: _____

Comments:_____

_____

_____

_____

_____

(Section 608 Inspection Checklist)

Level 2 Inspection Checklist (Section 608)
National Recycling and Emission Reduction Program

Facility Name:_____

Note: Complete the Level 1 inspection checklist first, and attach the level 2 checklist to it.

26. Reason for performing the Level 2 inspection?_____

_____
_____
_____

27. Is equipment owner certification paperwork in order?)i.e. forms)
    _____ Yes   _____ No

    If not, please explain:

    _____

    _____

    _____

    _____

Service Contractors
28. Are customer invoices available for inspection?
    _____ Yes   _____ No

29. Does the customer invoice record the type of service performed involving refrigerant?   _____ Yes   _____ No

30. Does the customer invoice provide the amount of refrigerant added to the appliance?   _____ Yes   _____ No

Scrap Recyclers, Landfills, or Other Disposal Facilities

31. Verification from appliance suppliers available for inspection?
    Yes _____   _____ No

32. What form of notification to appliance suppliers is provided?
    _____ Signs   _____ Letters to suppliers _____ Other _____

Owners/Operators of Commercial or Industrial Process Refrigeration Equipment

33. Are the service records or contractor invoices available for inspection?   _____ Yes   _____ No

34. Do service records or contractor invoices document leaks?
    _____ Yes _____ No

    Size of equipment charge _____ Rate of leak_____

35. Were leaks repaired within 30 days? ____ Yes ____ No

36. Does the facility have a dated retrofit or retirement plan?
____ Yes ____ No

37. Was the plan executed within one year of the plan date?
____ Yes ____ No

Facilities with Recovery Only Equipment

38. Are invoices or credit memos available for inspection verifying the return of recovered refrigerants? ____ Yes ____ No

39. Name of Refrigerant Reclaimer: _____

40. Reclaimer Address: _____

_____

41. Reclaimer Phone Number: _____

42. For any or all invoices or records reviewed, were photocopies made?
____ Yes ____ No (If yes, please attach photocopies)

Are invoices or other facility records:

___ Well-kept records with all information available?

___ Good records with only some missing required information?

___ Poor records with most lacking required information?

43. Inspector Name: _____

44. Inspector Signature: _____

45. Inspection Date: _____

46. Time of Inspection: _____

Comments: _____

_____

_____

_____

_____

(Level 2 Section 608 Inspection Checklist)

ENFORCEMENT GUIDANCE (SECTION 608)

Section 608(c)(1) of the Clean Air Act (CAA) states that "Effective June 14, 1993, it shall be unlawful for any person, in the course of maintaining, servicing, repairing, or disposing of an appliance or industrial process refrigeration, to knowingly vent or otherwise knowingly release or dispose of any Class I or Class II substance used as a refrigerant in such appliance (or industrial process refrigeration) in a manner which permits such substance to enter the environment. De minimis releases associated with good faith attempts to recapture and recycle or safely dispose of any such substance shall not be subject to the prohibition set forth in the preceding sentence."[1]

In addition, EPA is not proposing to consider refrigerant leaks from leaky equipment to be a violation of the venting prohibition, as an equipment leak is not a knowing release of refrigerant during service, maintenance, repair or disposal, but rather a release during the operation of the equipment. Neither will filling leaking equipment, or "topping-off" a system, be considered a violation, as in this instance, refrigerant is being added to equipment, not vented directly into the atmosphere. Also, Class I and Class II substances used as leak detection gasses and the use of these substances as holding charges in equipment that is being shipped are not covered by the venting prohibition.

Evidentiary Requirements

Evidence available to the Agency to prove a violation of this prohibition may include eyewitness testimony or an examination of records. These two options are discussed in detail below.

1) Eyewitness testimony

The EPA expects to receive tips from informants providing information that companies or individuals have vented refrigerants. The informant must be willing to sign an affidavit and to testify at an administrative hearing or trial regarding the venting incident. We cannot offer such an informant anonymity since his/her testimony will be the strongest evidence in any action taken. Such testimony can be supported by evidence gathered by Agency inspectors that there was no recovery or recycling equipment at the facility and records showing that the facility performed appliance repairs or disposed of appliances at the time of the alleged violation.

Where an informant is willing to testify that venting has taken place, EPA can issue a Finding of Violation or bring an administrative or judicial case. These enforcement options are discussed in more detail below.

---

[1]This section provides a summary of the EPA's Interim Enforcement Guidance for the National Recycling and Emission Reduction Program (Section 608). The complete text is available through the EPA TNN electronic Bulletin Board.

Awards Provision (Bounty)

This is a new provision which allows EPA to pay an award, not to exceed $10,000, to any person who furnishes information or services which lead to a criminal conviction or a judicial or administrative civil penalty assessed as a result of a violation of the CAA.

2) Other evidence

EPA may conclude that venting has taken place if a company has service invoices that state that service, maintenance or repair has been performed on appliances which contain refrigerant and if a company does not own or have access to recovery equipment. Some examples of the kinds of service which would imply venting are changing a compressor, replacing any coils, or changing filters.

Section 114 of the CAA gives EPA the authority to inspect records to determine compliance. If the region determines that sending a Section 114 letter requesting additional information is appropriate, the following questions may be useful in ascertaining whether a violation has taken place at a service facility:

"Have you or has your company serviced, repaired, maintained or disposed of any appliances containing refrigerants since July 1, 1992?"

"How many appliances containing refrigerants have you or has your company serviced, repaired, maintained or disposed of since July 1, 1992?"

"If you have or your company has serviced, repaired, maintained or disposed of any appliances containing refrigerants since July 1, 1992, what equipment did you or your company use to recover or recycle the refrigerants contained in the appliances?"

"If you or your company did not use recovery or recycling equipment in the course of servicing, repairing, maintaining or disposing of appliances containing refrigerants, what method did you or your company use to prevent the release of refrigerants into the atmosphere?"

Enforcement Options

If the Agency suspects that a violation of the prohibition on venting has taken place, EPA can pursue the following enforcement options:

1) Finding of Violation

EPA issues a Finding of Violation (FOV) letter if they have knowledge that a violation (i.e., knowing venting) has taken place in order to benefit from the presumption of continuing non-compliance[2]. An FOV can also be coupled with a request for additional information, as in a Section 114 letter.

2) Compliance Order

The EPA can order a person or business to take steps to return to compliance if a violation has taken place. Failure to comply with a compliance order is a violation of Section 113 of the Act and subject to penalties of up to $25,000. Some examples of orders EPA may issue in response to violations of the prohibition on venting are:

"You are ordered to cease releasing refrigerants into the environment during service, maintenance, repair, or disposal of appliances effective immediately."

"You are ordered to acquire equipment sufficient to avoid releasing refrigerants into the environment during service, maintenance, repair, or disposal of appliances effective immediately."

3) Administrative Cases

Violations can be addressed administratively under the authority of Section 113(d)(1) of the CAA. Cases cannot be brought administratively unless they involve penalties of less than $200,000 and the violations are less than twelve months old. Penalties will be calculated using the Interim Penalty Policy.

4) Civil Judicial Cases

If the violations involve penalties of more than $200,000 or occurred more than twelve months prior to filing, the cases must be referred to the Department of Justice.

5) Criminal Cases

If the evidence shows that the defendant not only knew he was releasing ozone depleting substances into the environment, but that he knew such an action was prohibited by law, the Agency should seek criminal penalties under Section 113(c) against the defendant.

---

[2] Section 113(e)(2) states, "where the Administrator or an air pollution control agency has notified the source of the violation, and the plaintiff makes a prima facie showing that the conduct or events giving rise to the violation are likely to have continued or recurred past the date of notice, the days of violation shall be presumed to include the date of such notice and each and every day thereafter until the violator establishes that continuous compliance has been achieved...."

Additional Considerations

The following information will be used to measure the seriousness of violations and in determining what type of enforcement action to pursue:

1) Number of venting incidents

This can be determined by an examination of service records or invoices, obtained either from the alleged violator using Section 114 authority, or from clients of the alleged violator.

2) Amount of refrigerant vented

It is EPA's position to assume a worst case environmental impact when assessing the gravity of a violation. Therefore, when determining the amount of refrigerant which may have been vented, we should assume that the entire charge of an appliance has been vented. For example, if the normal operating charge of an appliance is 25 kilograms of CFC 12, we can assume that the entire charge, all 25 kilograms, had been vented.

TAKING ENFORCEMENT ACTION

In determining whether to take enforcement action, the EPA may take into consideration whether a company has a purchase order for recovery equipment dated before July 1, 1992. Possession of a purchase order, however, cannot be construed to be compliance with the law, given the clear language of the Section 608(c).

In the case of an alleged violation, it is always appropriate to issue a Section 114 letter, requesting additional information in order to make a compliance determination. If sufficient evidence of a violation exists, EPA does have the authority to pursue both civil and criminal enforcement actions.

The EPA expects that enforcement actions will usually be taken against companies or businesses, not against individual technicians. EPA may enforce against individuals in appropriate circumstances, such as when conduct is knowing and willful or when an individual is a responsible company official.

CIVIL PENALTY POLICY (SECTION 608)

This section provides guidance for calculating the civil penalties EPA will require in pre-trial settlement of judicial enforcement actions, as well as the pleading and settlement of administrative enforcement actions, pursuant to Sections 113(b) and (d) and Section 608(c) of the Clean Air Act ("Act" or "CAA"), as amended, against persons who knowingly vent or otherwise knowingly release or dispose of any refrigerant in the course of maintaining, servicing, repairing, or disposing of an appliance.[3]

The Penalty for Violating the Act

Section 113 of the Clean Air Act allows EPA to seek penalties of up to $25,000 per day per violation. Each knowing venting or knowing releasing or disposing of refrigerant in a manner which permits such refrigerant to enter the environment in the course of maintaining, servicing, repairing, or disposing of an appliance constitutes a separate violation (each with a statutory maximum of $25,000).

EPA may in appropriate cases accept less than the statutory maximum in settlement. The penalty assessments contained reflect reductions from the statutory maximum which can be made based on the statutory penalty assessment criteria found in Section 113(e) of the Act.

Calculating a Penalty

In accordance with the general practice EPA follows when calculating all Clean Air Act civil penalties, penalties assessed for knowingly venting or otherwise knowingly releasing or disposing of any refrigerant in the course of maintaining, servicing, repairing, or disposing of an appliance will be the sum of an economic benefit component and a gravity component.

Economic Benefit

This component is a measure of the economic benefit accruing to the facility as a result of noncompliance with the Act. EPA will rely on the following matrix to determine the facility's economic benefit from delayed costs (failure to purchase equipment to comply with statute) and avoided costs (failure to properly operate and maintain equipment).

The matrix assumes that each facility that services small appliances would purchase one passive recovery device and all other facilities that service appliances would purchase one basic model of recovery equipment.

---

[3]Appendix X of EPA's Interim CAA Penalty Policy Applicable to Persons Who Maintain, Service, Repair, or Dispose of Appliances Containing Refrigerant.

ECONOMIC BENEFIT OF VENTING FROM APPLIANCES:

Where appliance contains:	Number of Months since Violation		
	1-3	4-6	7-9
< 1 pound of refrigerant	$28	$40	$60
> 1 pound but < 10 kgs	$65	$105	$150
10 kgs but < 50 kgs	$90	$130	$175
50 kgs but < 100 kgs	$120	$160	$205
100 kgs but < 500 kgs	$315	$355	$400
500 kgs or >	$540	$580	$625

Gravity

In addition to economic benefit, the violator must pay the gravity component of the penalty. The gravity component is the measure of the seriousness of the violation. The seriousness of the violation has two components: the importance to the statute and the potential environmental harm (ozone-depleting effect of the violator's actions) resulting from the violations.

Violations of the venting prohibition defeat the purpose of Section 608 by permitting the release of substances that degrade the stratospheric ozone layer. Their importance to the statutory scheme, therefore, requires the assessment of the following penalties:

A **penalty of $10,000** against all persons who knowingly vent or otherwise knowingly release or dispose of any stratospheric ozone depleting refrigerant in the course of maintaining, servicing, repairing, or disposing of an appliance and;

A **penalty of $15,000** against all persons who knowingly vent or otherwise knowingly release or dispose of any stratospheric ozone depleting refrigerant in the course of maintaining, servicing, repairing, or disposing of an appliance and who has previously been the subject of a Section 608(c) enforcement response (e.g. notice of violation, warning letter, administrative order, field citation, complaint, consent decree, consent agreement, or administrative or judicial order).

EPA acknowledges that the amount of ozone depleting refrigerant that enters the environment during a knowing venting, releasing, or disposing significantly affects the environmental harm resulting from the violations. The more refrigerant that is released, the greater the environmental harm. EPA, therefore, will assess an additional amount for each kilogram of refrigerant knowingly vented or otherwise knowingly released or disposed to ensure that the total penalty assessed appropriately reflects the seriousness of the defendant's violations.

EPA will assess \$60.00[4] for each kilogram of CFC-12 vented[5]. For all other refrigerants, EPA will multiply the ozone depletion potential (ODP) by \$60 to determine the per kilogram assessment for that substance. Section 602 of the Act lists all Class I and Class II ozone depleting substances and their ODPs, with the exception of two mixtures. CFC-500 is a mixture of CFC-12 (73.8%) and HFC-152a (26.2%) and has an ODP of 0.7. CFC-502 is a mixture of CFC-115 (51.2%) and HCFC-22 (48.8%) and has an ODP of 0.3. **This \$60 per kilogram amount is in addition to the \$10,000 assessment (or \$15,000 for subsequent violations) against the facility for violating Section 608(c).** Each violation of the venting prohibition constitutes a separate violation for which a separate penalty amount will be assessed. Because these violations are not of a continuing nature, no duration of the violation assessment attaches to these kinds of violations.

In keeping with the matrix provided by the Stationary Source Civil Penalty Policy, EPA will assess an additional amount to scale the penalty to the size of the violator. Adjustments to the gravity component must be made in accordance with the provisions of the Stationary Source Civil Penalty Policy.

Mitigating Penalty Amounts

Application of this policy significantly compromises the penalty amount EPA is authorized to pursue under the CAA. Penalty amounts calculated in accordance with this policy represent the minimum penalty that EPA can accept in settlement of cases of this nature. Reductions from this amount are acceptable only on the basis of the violator's demonstrated inability to pay the full amount (substantiated by the ABEL computer model) or other unique factors.

MVAC COMPLIANCE (SECTION 609)

Any service involving the motor vehicle air conditioner refrigerant requires the refrigerant to be recycled by certified technicians using certified equipment. Certified recycling equipment includes recover/recycle and recovery only equipment. Also, the sale of Class I and Class II substances, CFCs and HCFCs respectively, in containers less than 20 pounds is restricted to certified technicians. These containers are referred to as "small cans". The sales restriction on small cans is effective November 15, 1992. There is no restriction on the sale of containers weighing 20 pounds or more.[6]

---

[4] EPA has estimated that the benefit to be obtained from avoiding the release of 1 kilogram of ozone depleting substance ranges from \$60 - \$240 per kilogram of CFC-12. For the purposes of this penalty policy, it is assumed that the benefit is \$60/kg.

[5] CFC-12 has an ozone depletion wieght of 1.

[6] Reprinted from EPA guidance titled "Compliance Monitoring Strategy For Title VI Section 609 Servicing of Motor Vehicle Air Conditioners". File 609-CMS.WPF on EPA's TNN Bulletin Board.

In addition, motor vehicle repair shops, air conditioner service technicians, and small can distributors must comply with reporting and record keeping requirements under the Rule. Service shops are required to purchase refrigerant recycling or recovery equipment that has been certified by an EPA accredited laboratory. The purchasers of the equipment must certify to EPA that they have acquired certified recycling or recovery equipment, and that it is operated by properly trained and certified technicians. Technicians performing service must be certified to operate the equipment. To obtain certification a technician must complete an EPA accredited refrigerant recycling training course and pass the exam. The purchasers of small cans of refrigerant must present proof of certification. Distributors of small cans of refrigerant must certify in writing that the refrigerant is for resale with the end user being a certified technician. This written statement must be kept on file with the name of the purchaser and their business address.

While delegation of inspection authority is permitted under Section 609, the enforcement actions for violations of federal law can not be delegated under Section 609. Enforcement actions must be taken by EPA. The Compliance Strategy breaks down the three elements of the compliance program. They are education, compliance analysis, and compliance monitoring.

COMPLIANCE PROGRAM

There are five elements in the Compliance Program. To monitor compliance with these regulations EPA will monitor service entities, small can retailers, equipment owner certifications, technician certification courses, accredited laboratories, and substantially identical equipment. Inspections will be targeted at the service entities and small can retailers.

Service Entities and Small Can Retailers

There will be two levels of inspections for motor vehicle air conditioner service entities. A level 1 inspection will check for the presence of certified refrigerant recycling equipment and certified technicians. An in-depth, level 2 inspection will involve a more detailed records review. These records include the owners equipment certification, file copies of technician training, and if the business is using recover only equipment, records of the quantity of regulated refrigerant reclaimed off-site each month. Customer invoices should be requested and reviewed during a level 2 inspection. While the requirement to maintain these records was deleted from the Final Rule, inspectors may still request and use the invoices to make compliance determinations and as evidence in enforcement cases. There will also be inspections at small can retailers. These will be level 1 inspections. Inspectors will have to determine compliance by trying to purchase a small can of refrigerant without any proof of technician certification.

Equipment Owner Certification

The purchaser of certified recycling equipment must certify to EPA that approved recycling equipment has been purchased, and only properly trained and certified technicians will operate the equipment. The certification must also include the name, address, and telephone number of the service establishment; name of equipment manufacturer, model number, and serial number; and small entity certification if applicable. EPA has developed and distributed an example certification form. While not required, EPA encourages businesses to use the standard form because it makes the certifications easier to manage.

Technician Certification Courses

Technicians must be certified to operate refrigerant recycling equipment. Regulations require that technicians complete a course involving on-the-job training; training through self-study of instructional material; or on-site training involving instructors, video, or demonstrations. The technicians must learn the regulatory requirements, the proper use of refrigerant recycling equipment, and the environmental importance of recycling refrigerant. The technicians must pass an exam demonstrating their knowledge of the SAE J standards (J1989, J1990, J1991) and the recommended procedures for recycling air conditioning refrigerant. Future technology, and new refrigerant systems must also be addressed.

Laboratory Accreditation

EPA will accredit laboratories that wish to test and certify the refrigerant recycling equipment. Refrigerant recycling equipment must purify the air conditioner refrigerant according to SAE standard J1990 in order to be certified. Recovery only equipment must extract refrigerant in accordance with SAE standard J2209. This standard was proposed in a supplemental notice dated April 22, 1992.

Substantially Identical

In addition to monitoring certified refrigerant recycling equipment, SSCD will monitor substantially identical equipment. Uncertified equipment can become approved if it is determined to be "substantially identical" to equipment that is certified. Uncertified equipment must have been purchased prior to the proposal of the regulation. An example of circumstances under which equipment could be determined to be "substantially identical" may include equipment that was purchased from a manufacturer that no longer makes recycling equipment.

ENFORCEMENT GUIDANCE (Section 609)

Section 609 of the Clean Air Act Amendments of 1990 (the Act) regulates the servicing of motor vehicle air conditioners. The Act requires that any service performed on a motor vehicle air conditioner involving the refrigerant be performed by trained and certified technicians using approved equipment. In addition, the sale of refrigerant suitable for use in a motor vehicle air conditioner in containers of less than twenty pounds is restricted to certified technicians after November 15, 1992.

The Act does not give EPA the authority to delegate the enforcement of the section 609 regulation to state or local governments. Section 114 inspection authority, however, can be delegated to the states.

REGULATED COMMUNITY

Any establishment that services motor vehicle air conditioners for consideration or sells containers of refrigerant weighing less than twenty pounds is subject to this regulation.

VIOLATIONS

There are two categories of violations for the servicing of motor vehicle air conditioners and the sale of small cans of refrigerant; those that defeat the purpose of section 609, termed integrity of the regulation or substantive violations, and recordkeeping violations.

At service establishments violations include: failure to have and use approved recover/recycle or recover only equipment, and/or failure to have technicians properly trained and certified. Recordkeeping violations include: failure to keep records of where recovered refrigerant is sent off-site, failure to submit a small business certification and/or an equipment owner's certification, or failure to maintain or submit complete or accurate information.

Small can sales violations include: selling small cans to anyone other than a certified technician and failure to post a sign explaining the sales restriction. The recordkeeping violation is selling small cans to a distributor and not obtaining a signed statement that the cans are for resale only to a certified technician. To determine whether a small can retailer is in compliance with the regulations, an inspector must attempt to purchase a small can of refrigerant without showing or claiming to have technician certification. If the business does not make any attempt to request certification, and sells the small can(s) to the inspector, the business has violated the Act and the regulations.

Other violations include: an unapproved course issuing technician certifications, and an unaccredited laboratory approving equipment.

COMPLIANCE ACTIONS

Response to Noncompliance Tips

When reports are received about businesses that may not be in compliance with the regulation, a letter is sent to the alleged violator and the complainant, and an inspection should be scheduled.

Tip sheets are initiated that record the name, address, and owner of the business that is being reported; the name, address, and affiliation of the complainant, unless they wish to remain anonymous; the type of violation (recycling or small can); and the details of the complaint. Once a tip is received a "response to tip letter" is sent to the possible violator, a complainant thank you letter is sent, and an inspection is scheduled.

The "response to tip letter" notifies the establishment that the EPA has received information that they may not be in compliance. The letter identifies the possible penalties and states that the business may be inspected in the future. It includes the requirements of the Act and the regulations, and informs the business how to comply. In some cases it may be appropriate to combine this letter with a Section 114 request for information.

Inspections

Primarily service facilities and small can retailers will be inspected. The service facility inspections will predominately be level 1 inspections that check for approved equipment and certified technicians. There will also be a more in depth level 2 inspection at service facilities that will examine invoices. Level 1 inspections will be conducted at small can retail sales outlets because recordkeeping requirements for the sale of small cans of refrigerant are minimal.

CAA Section 114 Request for Information

A Section 114 request for information letter is sent to a service facility or small can retailer to gather information in order to determine compliance. Section 114 letters can be sent in response to a tip that an establishment is not in compliance when an inspection cannot be scheduled. A Section 114 letter can also be used after an inspection to gather information that may have been lacking in an inspection report, to discover possible additional violations, or to assess the extent of the violation for a possible enforcement action.

ENFORCEMENT ACTIONS

There are four types of enforcement responses; Administrative Order (AO), Administrative Penalty Order, Civil Judicial, and Criminal. Two types of letters that may be sent after an inspection or Section 114 request for information and prior to the issuance of an Administrative Penalty Order. They are a Finding of Violation (FOV) and an AO. In a section 609 enforcement action, an AO that includes an FOV will be appropriate in many cases.

Administrative Order

If the source remains in violation of the Act and the regulations the first enforcement response should be the issuance of an AO. The AO is appropriate for many MVAC cases because most violations can be corrected by an order to comply.

A FOV notifies the party that they are in violation. The Agency is not required to issue FOV's for violations of Section 609. It is common practice in some regions to use a FOV alone, but we recommend it in combination with an AO. The AO will describe the finding of violation(s) and order the establishment to comply with the regulations. The establishment can come into compliance by having their technicians trained and certified by an EPA approved training course, purchasing approved refrigerant recover/recycling or recover only equipment, or stopping the sale of small cans of refrigerant to anyone other than certified technicians. A finding of violation letter may be modified to order the facility to correct the violation.

CAA Section 113(d): Administrative Penalty Order

Administrative Penalty Orders will probably be the most commonly used enforcement response for MVAC cases. Administrative penalty cases are initiated with an Administrative

Penalty Order. To assess administrative penalties, the date of the first alleged violation must be less than 12 months from initiation of the enforcement action and the penalty assessed must be less than $200,000. Administrative penalties are preferable because they make for faster resolution of cases and are less resource intensive for both the Agency and the respondent than civil judicial proceedings.

CAA Section 113(b): Civil Judicial Actions

Civil judicial enforcement actions can be taken in any case as long as it is less than five years from the date of the alleged violation. Civil judicial cases must be brought if the penalties sought exceed $200,000, or the initiation of the enforcement action is more than 12 months from the first alleged date of violation, or obtaining compliance will take more than one year, and substantial court-supervised injunctive relief is required. Civil judicial actions may require extensive post filing discovery. Section 113(b) authority may be preceded by pre-referral negotiations. See Clean Air Act Compliance/Enforcement Guidance Manual, revised July 1987. Also see Process for Conducting Pre-Referral Settlement Negotiations on Civil Judicial Enforcement Cases, April 13, 1988. Because the MVAC penalties will not usually exceed $200,000 and compliance can be achieved relatively easily, it will be rare for this type of action to be taken.

CAA Section 113(c): Criminal Actions

To bring a criminal case, the defendant must have either knowingly committed the violation or knowingly given a false statement or certification. Due to the difficulty in proving an intentional violation, few criminal cases are expected.

CHAPTER ELEVEN

RECORD KEEPING & REPORTS

— REQUIREMENTS FOR EVERYONE —

Refrigeration businesses, technicians, owner/operators of certain refrigeration equipment, manufacturers of recovery/recycling equipment, wholesalers, reclaimers, disposers and recovery/recycling equipment owners are required by the EPA to maintain certain records. EPA originally had a long list of record keeping requirements but much of that was removed when section 608 and 609 final rules were published. This chapter lists many of the records and reports that EPA requires for sections 608, 609 Motor Vehicle Air Conditioning, section 611 Labeling, and DOT 49CFR transportation. The majority of records must be retained for 3 years. Disposers must retain the records on appliances at the site. References are included so that you may easily refer to the rules in the appendix of this book. Additional actions will be identified with this symbol (✲) that will help aid your record keeping efforts and assist you and your shop to prepare for EPA inspection.

NOTE: Individual states may have additional record keeping and reporting requirements. This information is ONLY for FEDERAL EPA requirements.

SERVICE TECHNICIANS TAKE NOTE: failure on your part to comply with these requirements can result in your employer being fined by EPA, OR the EPA inspector can levy the fine against YOU!

RECOVERY/RECYCLING EQUIPMENT OWNERS

Section 608, 82-162, requires that persons maintaining, servicing, or repairing appliances (except MVACs), and persons disposing of appliances (except small appliances, MVACs and MVAC-like equipment) must certify that such person has acquired and is properly using certified recovery or recycling equipment. This must be done within 20 days of commencing business. The following information must be recorded:

1. The name and address of the purchaser of the equipment, including the county name.

2. The name and the address of the establishment where each piece of equipment is or will be located.

3. The number of service trucks (or other vehicles) used to transport technicians and equipment between the job sites and establishment and the field.

4. The manufacturer's name, the date of manufacture, and if applicable, the model and serial number of the equipment; and

5. A statement that the equipment will be properly used in servicing or disposing of appliances and that the information given is true and correct. Refer to CFR 82-162 for the location to mail your certification.

Section 609

Recovery/recycling equipment that you use must be registered with EPA whether you own it, lease it or rent it for a day. For section 609 the form has a statement that whoever signs the form attests that anyone who uses the equipment is a trained and CERTIFIED technician.

RECORDKEEPING SECTION 608, CFR 82-166

Records that are required to be maintained by this section must be kept for a minimum of 3 years unless otherwise indicated. *ANYONE* who *SELLS* or *DISTRIBUTES* any Class I or II substance for use as a refrigerant MUST retain invoices with the name of the purchaser, date of sale, and the quantity of refrigerant purchased.

❋ For each service call the invoice should identify the equipment, the date, amount of refrigerant added, service performed and the name, phone number and address of the owner. As you consume refrigerant on service calls you have to purchase more. Keep records of your purchases from wholesale houses. Some wholesale houses will send you a monthly invoice of refrigerant purchases that will help you maintain accurate purchase records for EPA inspections. Purchasers of Class I or II refrigerants who employ technicians that recover refrigerants *MAY* provide evidence of each technician's certification to the seller of the refrigerant. The seller will then keep this information on file. The purchaser is required to notify the seller of any change in a technician's employment or certification status. SECTION 608 CERTIFIED TECHNICIANS ARE NOT AUTHORIZED TO PURCHASE "SMALL CANS" AS IDENTIFIED IN SECTION 609, 82-34.b.

Reference CFR 82-166.(i), Persons disposing of small appliances, MVACs and MVAC-like appliances must maintain copies of signed statements per CFR 82-156(f)(2). The statement verifies that the refrigerant has been evacuated from the appliance previously. These statements are to be SIGNED by the person from whom the appliance or shipment is obtained and attests that *ALL* refrigerant that had not leaked has been recovered in accordance with CFR 82-156 (g) or (h). The name and address of the person who did the recovery and the date of the recovery to be provided, or a contract with the supplier that states all refrigerant will be removed before delivery. The contract does not bypass the records requirements stated above. Persons involved in the final step of the disposal process must notify suppliers of appliances that refrigerant must be properly removed BEFORE delivery to the facility (land fill) for disposal. Warning may be by signs, letters to suppliers or other equivalent means.

EQUIPMENT TESTING ORGANIZATIONS

Approved testing organizations MUST maintain records of equipment testing and performance and a list of equipment that meets EPA requirements. Within 30 days of EPA approval as an equipment testing organization, a list of all certified equipment shall be submitted to EPA and annually at the end of each calendar year thereafter. Approved testing organizations shall submit to EPA within 30 days of the certification of a new model line of recovery or recycling equipment, the name of the manufacturer and the name and/or serial number of the model line. Certification may be revoked for recovery/recycling equipment that fails a retest or if an inspection of the manufacturing facility shows that a previously certified model line fails to meet EPA requirements. The approved testing/inspecting organization must send the results to the EPA and they must be received within 30 days of the retest or inspection. Equipment manufacturers or importers of recovery/recycling equipment must have their recovery/recycling equipment either retested or the manufacturing facility inspected to ensure that each approved equipment model line continues to meet the required certification criteria. This must be done every 3 years from the date of initial certification, CFR 82-158. Failure to meet the certification criteria established for an equipment model line is cause for revoking that equipments' certification. The equipment manufacturer or importer shall be notified of the revocation and the basis for the determination by the EPA administrator or designated representative.

Equipment certification involves more than just meeting evacuation requirements; it also attests to the fact that the manufacturer continues to make certified recovery/recycling equipment using components of the same quality and design that were used in the original unit tested and certified.

SECTION 609 RECORD KEEPING, CFR 82-42

Anyone servicing or repairing MVAC equipment with refrigerant must be EPA certified and use approved equipment, effective date August 13, 1992.

As of January 1, 1993, any person or business repairing or servicing MVACs for consideration SHALL certify: that approved recovery/recycling equipment has been acquired, and is being used properly, that each individual authorized to use the equipment is properly trained and certified; the name of the purchaser; the address where the equipment will be located; and the manufacturers name and equipment model number, date of manufacture and serial number and that the information is true and correct. If the recovery/recycling equipment or business is sold, the buyer has 30 days to refile the certificate of compliance. Appendix V includes the form.

The statement "properly trained and certified" as stated in section 609 means that the person has met the requirements of a technician certification program per CFR 82-40, and was trained on the proper operation of owned recovery/recycling equipment. As an example, to document the proper use of approved recovery/reclaiming equipment type up a letter or training form (see appendix VIII) that says "Mr. Jones received 2 hours of training on proper use of FREDY FREON Model A recovery equipment on July 4, 1994, conducted by Mr. Sam Refrigeration. Mr. Jones passed the training." This documentation confirms that the technician had training and at that time knew how to properly use the equipment. There is no EPA requirement for this letter or training form but it is an easy way to record that successful training was given on the recovery equipment and who received it. ✸ Mr. Refrigeration must be EPA certified to use the equipment for actual demonstrations.

Any person who owns approved refrigerant recycling equipment must maintain records of the names and addresses of any facility to which refrigerant is sent. Persons owning approved refrigerant recycling equipment must retain records demonstrating that all persons authorized to operate the equipment are currently certified under CFR 82-40. If someone purchases Class I or II refrigerants from you for resale only, the seller must obtain a written statement from the purchaser that the containers are for resale only and indicate the purchasers name and business address. Anyone selling Class I or II refrigerants that may be used in a MVAC in containers less than 20 lbs MUST prominently display a sign at the point of sale on the prohibition of selling to non-certified personnel, CFR 82-42(c). TECHNICIANS CERTIFIED UNDER SECTION 608 ARE NOT AUTHORIZED TO PURCHASE "SMALL CANS". Unless otherwise noted all records are to be maintained for 3 years. Organizations that service MVACs must keep these records on site. Upon presenting credentials authorized EPA inspectors must be given access to all records that are required and maintained at the facility.

WHOLESALERS / RETAILERS / DISTRIBUTORS

These groups must maintain usual business records of their refrigeration transactions, including the name of the buyer, date and the quantity sold. They must verify that the purchaser is properly trained and certified and must have a reasonable basis to believe the information presented by the purchaser is accurate. A sign must be displayed at the point of sale of small containers stating that it is a violation of federal law to sell containers of Class I and Class II refrigerant in containers of less than 20 pounds to anyone who is not properly trained and certified. NOTE: 608 certified technicians are NOT AUTHORIZED to make small can purchases! ✱ You also need to keep records of refrigerant sent to be reclaimed/reprocessed, where it went, when it went, who sent it and how much. ✱ I would also record the buyer's EPA number and business address for my records for ANY sale of small cans.

For Section 611 labeling, CFR 82-118, You are required to pass through the labeling information that accompanies the product. CFR 82-124 requires a warning statement if the container or product is introduced into interstate commerce.

RECLAIMERS

Reclaimers must maintain records of the names and addresses of PERSONS sending them materials for reclamation and the quantity of the material (the combined mass of refrigerant and contaminants) sent to them. These records will be kept on a transaction basis. On an annual basis, reclaimers are required to keep records of the mass of material sent to them, the mass of refrigerant reclaimed and the mass of waste products. This is to be reported to the administrator annually within 30 days of the end of the calendar year. These records allow determination of refrigerant releases from processing. CFR 82-164 lists these requirements to become a certified reclaimer: If reclaiming for sale to a new owner you must certify to the EPA administrator that refrigerant will be returned to at least the standard of purity set by ARI Standard 700-1988, Specifications for Fluorocarbon Refrigerants. Verify the purity using the methods set forth in ARI standard 700-1988. Release no more than 1.5% of the refrigerant during the reclamation process and dispose of wastes from the reclamation process in accordance with all applicable laws and regulations and that the information given is true and correct. A signature is required by the owner or responsible officer of the reclaimer. A change of ownership requires a NEW certificate and it SHALL be accomplished within 30 days of the change of ownership. Section 611, CFR 82-108, 82-110, 82-114 and 82-124 apply.

DISPOSERS

Persons disposing of small appliances, MVACs, and MVAC-like appliances must maintain copies of signed statements obtained in accordance with 82-156(f)(2). Statements must have the name and address of the person who recovered the refrigerant and the date recovered. The reference provides specific verbiage requirements. These statements verify that the refrigerant has been evacuated from the appliance(s) previously and they allow the disposer to show that proper evacuation occurred at a point before reaching the disposer, thus shielding the disposer from liability. Persons who knowingly provide false statements will be subject to CRIMINAL penalties.

OWNER/OPERATORS of AIR CONDITIONING & REFRIGERATION EQUIPMENT

Owner/operators who own appliances that contain 50 or more pounds of refrigerant must maintain records of servicing and amounts of refrigerant added to machines. These records may include service invoices provided by the technician. Records must also be kept by the owner/operator of refrigerant purchased and added to equipment where owners add their own refrigerant. These documents assist owners and the technician in determining if leaks exist and the rate of leakage. Leak rate determines the need for repairs, retrofit or retirement of the appliance. All documents must show the date any refrigerant was added.

TECHNICIAN CERTIFICATION PROGRAMS

Technician certification programs must retain names and addresses of all individuals taking tests, the scores of all certification tests administered, and the dates and locations of all tests administered. They must send EPA an activity report every six months to include the pass/fail rate and testing schedules. Any test bank question that needs to be modified should be included with the report with information about the question. Each approved certifying program must display a copy of the approval letter from EPA at each testing center. A test may be administered orally to any person who makes this request, in writing, to the program at least 30 days before the scheduled date for the examination. The letter must explain why the request is being made. Upon successfully obtaining certification, individuals must be issued a wallet-sized card to be used as proof of certification. Successful completion of the closed book test requires a score of at least 70% or higher. If a TYPE I certification mail-in format test is used a score of 84% is required to pass the exam. Each certification card must include, at a minimum, the name of the certifying program, the date of program approval, the name of the person certified, the TYPE of certification, a unique number for the certified person, and the following statement: (Name of person) has been certified as a (Type I, Type II, Type III, and/or Universal, as appropriate) technician as required by 40CFR part 82, Subpart F. Technicians must keep a copy of their certification at their place of business. More information on the standards for becoming a testing program as well as EPA inspection of security and testing may be found in Section 608, Appendix D to Subpart F.

LABELING 40 CFR Part 82, Subpart E, Section 611

No specified record keeping or reports are required for CFC or HCFC refrigerants. All containers in which a Class I or Class II substance is stored or transported and all products containing a Class I substance or Class II substance require labeling. WARNING STATEMENT REQUIREMENTS. Each container or product identified in CFR 82-102(a) or (b) shall bear the following warning statement, meeting the requirements of this Subpart for placement and form:

WARNING: Contains (or Manufactured with, if applicable) (insert name of substance), a substance which harms public health and environment by destroying ozone in the upper atmosphere.

The warning statement shall not interfere with, detract from, or mar any labeling information required on the label by federal or state law. PLACEMENT of the warning label is such that the warning statement must be clearly legible and conspicuous and it appears with such prominence and conspicuousness as to render it likely to be read and understood by consumers under normal conditions of purchase. ❋ To get an idea of what this looks like look at any disposable jug and the box it came in. Both will have the warning statement on them as long as they were intended for interstate commerce. CFR 82-108 presents variations of how the warning statement may be affixed. ❋ When you transport recovery vessels containing recovered refrigerant to the shop or wholesale house for recycling they must be marked with the warning statement. There may also be state and local laws that require labeling and, of course, federal DOT has requirements. Check with your wholesale house or recycler for warning placards, tags or adhesive-backed warning labels to mark your tanks. Make sure any labels meet the requirements of CFR 82-110 for color, size of lettering and format.

RECORD KEEPING TIPS

1.   Post the EPA HOTLINE phone number.

2.   Discard your old service forms and have new ones printed that have required information clearly marked.

3.   The form should have: Your company name, Date block, Name of owner of the equipment, Address of service call, Phone # of owner, Type of equipment, Manufacturer, Type of refrigerant the equipment uses, How many pounds added to appliance for charging, Services performed, Problem found, A block to indicate if leak testing was done, Leak rate percentage calculation (if required), Recommended factory charge (in pounds and ounces), Pounds of recovered refrigerant. If you reuse a lot of recovered refrigerant, consider a separate authorization form for signature or initials of owner showing they authorized you to reuse their recovered refrigerant, with a disclaimer to not hold you liable for damages.

4.   Establish a written procedure for disposal of recovered refrigerants. Try to find a wholesale house that automatically sends you a monthly report of quantities of refrigerants returned and purchased. This is a key inspection point.

5.   EPA requires a copy of the technician type certification be at the shop so show some pride, buy a frame and hang it on the shop wall.

6.   Make sure you have MSDS sheets for all the materials that require an MSDS sheet. Keep the forms in a bright color binder and make sure everyone knows where it is.

7.   Any time you register recovery/recycling equipment or add a new technician to an equipments certification, send the new or updated certification by certified mail. Your receipt and copy of the registration is your evidence of compliance.

8.   Always document training given technicians on recovery/recycling equipment. Record: date, time, where, how long, who attended, what it is on or about and who presented the training. See Appendix VIII for more training information.

9.   Make up a check list for testing your recovery/recycling equipment daily for leaks and other factory recommended tests.

10. Establish service requirements and follow them for each piece of recovery/recycling equipment.

11. Ensure there are no disposable jugs being reused in trucks or the shop.

12. Review established policy with all personnel so they are well aware of what to do or where to look for guidance and the necessity of their compliance and your support.

APPENDIX I

ANSWERS TO PRACTICE QUESTIONS

This Appendix lists the answers to the sample questions from Chapters Six through Nine. The answers to the closed book practice exam in Appendix III follow the exam in that section. Answer the questions first, then after reading the appropriate references in this book, check your answer against the correct answers listed below.

If you have difficulty with any one section, review the appropriate reference material in this guide. If you have difficulty with air conditioning and/or refrigeration theory or system operation refer to basic theory references such as **The Refrigeration Service Engineers Society's (RSES)** *Service Application Manual* (SAM). This four volume set is an excellent reference and it may be available at a technical library near you. If your shop doesn't have this reference it can be obtained from RSES, 1666 Rand Rd., Des Plaines, IL 60016-3552, They can be reached by phone at (708) 297-6464. Business News Publishing Company, Technical Book Division, P.O. Box 2600, Troy, MI 48007-2600, phone 1-800-837-1037, also publishes HVAC/R references including *Refrigeration Fundamentals* by William Gorman or *Basic Refrigeration* by Guy King.

It is also important to ensure that your shop has the latest EPA updates for rule changes and final decisions. Also, SNAP - Significant New Alternatives Policy - approves alternative refrigerants and they update their lists of acceptable and unacceptable substitutes as new substitutes are tested and approved. Stay in contact with the EPA through the many resources listed in this book and use their electronic bulletin board and toll free numbers to insure you have the latest available information.

CHAPTER SIX
SECTION 608 GROUP 1 ANSWERS

1. D	21. D	41. A	61. D
2. C	22. A	42. D	62. D
3. C	23. C	43. D	63. D
4. A	24. B	44. A	64. D
5. C	25. D	45. A	65. C
6. D	26. A	46. A	66. D
7. N	27. C	47. D	67. A
8. D	28. A	48. B	68. B
9. C	29. A	49. C	69. A
10. C	30. C	50. D	70. C
11. D	31. A	51. D	71. D
12. D	32. D	52. B	72. B
13. B	33. D	53. B	73. A
14. B	34. B	54. B	74. C
15. A	35. C	55. A&C	75. C
16. B	36. A	56. D	76. C
17. C	37. D	57. B	77. C
18. B	38. A	58. C	78. C
19. A	39. C	59. D	79. B
20. C	40. B	60. C	80. D

CHAPTER SEVEN
SECTION 608 ANSWERS GROUP II
TYPE I QUESTIONS

1. C	14. C	27. A	39. B
2. B	15. B	28. B	40. A
3. C	16. D	29. B	41. C
4. A	17. D	30. D	42. A
5. C	18. A	31. B	43. D
6. B	19. D	31. D	44. D
7. C	20. B	32. C	45. B
8. A	21. D	33. B	46. D
9. B	22. B	34. A	47. A
10. D	23. D	35. A	48. A
11. C	24. C	36. C	49. D
12. B	25. D	37. A	50. B
13. B	26. D	38. B	51. D

TYPE II QUESTIONS (Page 57)

1. C	12. A	23. B	33. C
2. C	13. C	24. D	34. B
3. B	14. D	25. B	35. D
4. C	15. D	26. C	36. C
5. C	16. D	27. D	37. A
6. D	17. C	28. A	38. D
7. D	18. C	29. B	39. D
8. C	19. D	30. C	40. A
9. D	20. A	31. B	41. D
10. A	21. D	32. C	42. C
11. B	22. A		

TYPE III QUESTIONS (Page 60)

1. C	3. A	5. C	7. B
2. C	4. A	6. A	

CHAPTER 8
MVAC SECTION 609 ANSWERS

1. C	10. D	19. C	28. B
2. B	11. B	20. B	29. C
3. C	12. C	21. B	30. B
4. A	13. B	22. C	31. A
5. B	14. A	23. C	32. C
6. A	15. B	24. B	33. C
7. C	16. A	25. B	34. B
8. B	17. D	26. D	35. A
9. B	18. D	27. B	36. D

CHAPTER 9
RECOVERY/RECYCLING/RECLAIMING 608/609

1. B	15. D	29. B	43. C
2. C	16. C	30. C	44. D
3. A	17. A	31. B	45. B
4. B	18. A	32. B	46. C
5. B	19. C	33. B	47. B
6. D	20. D	34. C	48. B
7. D	21. A	35. A	49. A
8. B	22. C	36. A	50. B
9. C	23. D	37. D	51. A
10. A	24. A	38. C	52. B
11. D	25. D	39. A	53. A
12. A	26. B	40. D	54. C
13. C	27. B	41. A	55. A
14. D	28. B	42. C	

NOTE: The answers to the closed book practice exam in Appendix III follow the exam.

APPENDIX II

EPA APPROVED TESTING
SECTION 608/609

The following lists of EPA technician certification programs will be updated by the EPA when other technician certification programs are approved. Programs appearing on this list are approved to provide the EPA technician certification test; however, this approval does not address the status of technicians that participated in any *voluntary* training and/or certification program. Issues concerning previously trained technicians will be announced as programs are approved. In addition, EPA does not review or approve any training programs or materials. Please note: The addresses and phone numbers listed below are for the programs' headquarters. Many programs offer testing locations throughout the country. Each program will be able to provide you with their testing locations.

These lists are compiled from EPA publications. Official copies of this material and future updates can be obtained from the Government Printing Office (GPO) at 202-512-1530 or by calling the EPA Hotline at 1-800-296-1996. If you have a personal computer with a modem and a communications program you can download various updates from the EPA's TTN Clean Air Act electronic bulletin board. The access number is 919-541-5742. You can also access the **FEDWORLD National Technical Information Service (NTIS)** electronic bulletin broad. From FEDWORLD you can access more than 100 federal bulletin board systems operated by the U.S. government, including EPA's Office of Research & Development. To connect to FEDWORLD, use your computer communication software to dial FEDWORLD at 703/321-8020. Set your parity to NONE, Data Bits to 8 and Stop Bit to 1 (N,8,1). Set your terminal emulation to ANSI or VT-100. After making a connection access bulletin board # 94 (ORDBBS (EPA).

SECTION 608 TECHNICIAN CERTIFICATION PROGRAMS

AC/C Tech
4415 N. Forest Manor Avenue
Indianapolis, IN 46226-3080
Phone: 317-545-7071

Air Conditioning Contractors of America
ACCA/Ferris State University (FSU)
1513 16th St.
Washington, DC 20036
Phone: 202-483-9370

Air-Conditioning & Refrigeration Inst ARI
4301 North Fairfax Drive Ste 425
Arlington, VA 22203
Phone: 703-524-8800

Delaware County Community College
Route 252 & Media Line Rd.
Media, PA 19063
Phone: 215-359-5338

Environmental Training Group Inc.
9677-3 Basket Ring Road
Columbia, MD 21045
Phone: 410-740-1551

General Services Administration
(Reserved for federal employees)
18th & F St. NW Rm 4318
Washington, DC 20405
Phone: 202-501-0429

National Assessment Institute
Block & Assoc. (NAI/Block)
5700 SW 34th St. Ste. 1303
Gainesville, FL 32608
Phone: 904-373-8421

National Assoc. of Power Engineers (NAPE)
5-7 Springfield St.
Chicopee, MA 01013
Phone: 413-592-6273

National Association of Plumbing
Heating & Cooling Contractors (NAPHCC)
P.O. Box 6808
Falls Church, VA 22040
Phone: 800-533-7694

New York City Transit Authority (NYCTA)
Reserved for Transit Authority Personnel
2125 West 13th St.
Brooklyn, NY 11223
Phone: 718-265-4267

North Carolina State Board
Refrigeration Examiners
P.O. Box 30693
Raleigh, NC 27622
Phone: 919-781-1602

Operating & Maintenance Engineering
Trade Trust Fund of California & NV (OME)
2501 W. Third St.
Los Angeles, CA 90057
Phone: 213-385-2889

Refrigeration Environmental Protection
Association (REPA)
7525-M Connelley Drive
Hanover, MD 21076
Phone: 800-435-3331

Refrigeration Service Engineers Society (RSES)
1666 Rand Road
Des Moines, IL 60016-3552
Phone: 708-297-6464

Sequoia Institute
420 Whitney Place
Fremont, CA 94539
Phone: 510-490-6900

Texas Engineering Extension Service
The Texas A&M University
9350 South Presa
San Antonio, TX 78223-4799
Phone: 210-633-1000

The Refrigeration School Inc. (RSI)
4210 East Washington St.
Phoenix, AZ 85034-1894
Phone: 602-275-7133

United Assoc of Journeymen & Apprentices of the Plumbing/Pipe Fitting Industry US/CA
901 Massachusetts Ave. NW
Washington, DC 20001
Phone: 202-608-5823

Universal Technical Institute (UTI)
3002 North 27th Ave.
Phoenix, AZ 85017
Phone: 602-271-4174

University Technical Institute (UTI)
902 Capitol Ave.
Omaha, NE 68102-9954
Phone: 402-345-2422

VGI Training Div (VGI)
Video General Inc.
1156 107th St.
Arlington, TX 76011
Phone: 800-886-4109

SECTION 609 TECHNICIAN CERTIFICATION PROGRAMS

CFC Reclamation & Recycling Service Inc.
P.O. Box 560
Abilene, TX 79604
Phone: 915-675-5311

Geneva Steel
P.O. Box 2500
Provo, UT 84603
Phone: 801-227-9000

Greater Cleveland Auto Dealers Assoc.
6100 Rockside Woods Blvd Ste 235
Independence, OH 44131
Phone: 216-328-1500

Int'l Mobile Air Conditioning Association
P.O. Box 9000
Fort Worth, TX 76147-2000
Phone: 817-338-1100

Jiffy Lube International
P.O. Box 2967
Houston, TX 77252-2967
Phone: 713-546-4100

K-Mart Corporation East/Central Regional Auto Training Center
2300-B West Higgins Road
Hoffman Estates, IL 60195-2491
Phone: 708-884-1146

Mechanic's Education Association
1805 Springfield Avenue
Maplewood, NJ 07040-2910
Phone: 201-763-0086

Mobile Air Conditioning Society
P.O. Box 97
East Greenville, PA 18041
Phone: 215-541-4500

National Institute of
Automotive Excellence
13505 Dulles Technology Drive
Herndon, VA 22071-3415
Phone: 703-713-3800

New York State Association of
Service Station & Repair Shops Inc.
8 Elk Street
Albany, NY 12207
Phone: 518-434-6102

Potomac Electric Power Company
8400-B Old Marlboro Pike
Upper Marlboro, MD 20772
Phone: 301-967-5294

Rancho Santiago College
17th at Bristol St.
Santa Ana, CA 92706
Phone: 714-564-6661

Ryder Truck Rental Inc.
3600 NW 82nd Avenue
Miami, FL 33102-0816
Phone: 305-593-3684

Snap-on Tools Corporation
2801 80th St.
Kenosha, WI 53141-1410
Phone: 414-656-5200

Texas Engineering Extension Service
San Antonio Training Division
9350 South Presa
San Antonio, TX 78223-4799
Phone: 512-633-1000

U.S. Army Ordnance Center & School
ATTN: ATSL-DT-TDS-TS
Aberdeen Proving Grounds
Aberdeen, MD 21005-5201
Phone: 410-278-4099

Waco Chemicals Inc.
12306 Montaque St.
Pacoima, CA 91331
Phone: 818-897-3018

Wayne Supply Company
P.O. Box 35900
Louisville, KY 40323-5900
Phone: 502-774-4441

Yellow Freight System Inc.
10990 Roe Avenue
Overland Park, KS 66207
Phone: 913-345-3000

APPENDIX III

CLOSED BOOK PRACTICE EXAM

The closed book practice exam questions are presented in several formats to help you retain the information. Select the most correct answer for each question. The answers to this test are included at the end of this Appendix.

If you need to brush up on air conditioning and/or refrigeration theory or system operation refer to references such as **The Refrigeration Service Engineers Society's (RSES)** *Service Application Manual* (SAM). This four volume set may be available at technical libraries or you can order it from RSES, 1666 Rand Rd., Des Plaines, IL 60016-3552, phone (708) 297-6464. Business News Publishing Company, Technical Book Division, P.O. Box 2600, Troy, MI 48007-2600, phone 800-837-1037, also publishes HVAC/R references, including *Refrigeration Fundamentals* by William Gorman and *Basic Refrigeration* by Guy King.

TEST TAKING STRATEGIES

The following strategies will help you improve your grades. Use these strategies on the practice tests in this book, on the closed book practice exam in Appendix III, and when you take your actual EPA certification exam. If you practice these techniques now, when you take the certification exam they will become second nature.

✎   Eliminate the answers in multiple choice questions that make no sense at all. You can often eliminate half of the answers through this method. If you have to guess an answer, you improve your chances through the process of elimination.

✎   Be skeptical when an answer includes words like, always, never, all, none, generally, or only. These words can be a trap. Only select an answer with these words in it if you are absolutely sure it is the right answer.

✎   If two answers have opposite meanings, look closer. Many times one of the two is correct.

✎   Place a mark next to answers that you are unsure about. After completing the remainder of the exam, go back and review these questions and make a final selection. Often, other questions that you've answered will jog your memory.

✎   One word can dramatically change the meaning of a sentence. Read each question word-for-word before answering.

✎   Don't let the test get the best of you. Build your confidence by answering the questions you know first. If the first question you read stumps you, skip it and go on to the next one. When you've completed most of the exam you can go back - if time permits - to the questions that you couldn't answer.

✎   Get plenty of rest the night before the exam.

GROUP I SAMPLE TEST
(25 questions)

1. The restrictions on opening appliances, recovering refrigerant, disposal of appliances and selling used class I or II refrigerants begin

    a. August 12, 1993
    b. November 15, 1993
    c. June 14, 1993
    d. July 13, 1993

2. When broken down by solar radiation the chemical element released by CFCs that damages ozone is

    a. hydrogen
    b. oxygen
    c. triatomic oxygen
    •d. chlorine

3. Agroscopic means

    a. acidity
    b. absorbs water
    c. alkaline
    d. base chlorides

4. What does the term "drop in substitute" mean?

    a. A substitute refrigerant that does not require modification of existing appliance.
    b. A substitute refrigerant that requires some modification to the appliance.
    c. A substitute refrigerant like HFC-134A.
    d. A refrigerant that may be mixed with existing appliance refrigerant.

5. The penalty for violations of section 608 are

   a. up to a $25,000. fine.
   b. misdemeanors up to $500 per day fine.
   c. up to $25,000 per day, per occurrence.
   d. $2,500 maximum per day for all occurrences.

6. What is the proper attire for working with R-22?

   a. Safety goggles and butyl-lined gloves.
   b. Safety goggles and leather gloves.
   c. Safety goggles.
   d. Safety goggles and latex gloves.

7. A storage or recovery vessel must be

   a. UL or DOT approved.
   b. UL approved.
   c. filled to rated capacity.
   d. pressure tested every 4 years.

8. Unfavorable chemical breakdown of refrigerants is usually caused by

   a. friction.
   b. freon burn.
   c. hydrofluoric acid.
   d. water.

9. The only source of CFC-12 after production stops is

   a. recovery.   b. importing.
   c. recycling.   d. disposal.

10. Refrigerant blends that do not use mineral oils use

    a. benzacol oils.
    b. polyalphitic oils.
    c. aldehyde resins.
    d. alkylbenzenes.

11. Before converting equipment to another refrigerant ensure

    a. that there are no leaks.
    b. compatibility with the new refrigerant.
    c. that the manufacturer approves.
    d. (a), (b), and (c) are correct.

12. The mixing of three different refrigerants that exhibit separate characteristics under load, though considered one refrigerant is a, (an)

    a. thermal blend.
    b. azeotrope.
    c. ternary blend.
    d. (b) and (c)

13. Running a compressor with a deep vacuum will cause

    a. sticking valves.
    b. burn out.
    c. brown out.
    d. sublimation.

14. HFC refrigerants may use what oil(s)?
    a. alkylbenzenes.
    b. polyalkaline glycols.
    c. polyolesters.
    d. (b) and (c).

15. The appliance group most responsible for CFC emissions are

    a. small appliances.
    b. purge units.
    c. chillers.
    d. MVAC and MVAC-like appliances.

16. Synthetic oils' exposure to air must be minimized because these oils are

    a. miscible.
    b. aldehydes.
    c. hygroscopic.
    d. water based.

17. ASHRAE standard 15 requires that an equipment room using B1 refrigerant

    a. contain refrigerant vapor sensors.
    b. contain fire suppression sensors.
    c. have exhaust vent fans.
    d. have both (a) and (c).

18. What lubricant will clean pipe scale, sludge, and other contaminates from appliances?

    a. Mineral oil   b. Benzacol oil
    c. Polyolester oil   d. Aldehyde oil

19. A refrigerant having more than one temperature vs pressure curve is said to exhibit

    a. break down.
    b. freon burn.
    c. temperature glide.
    d. non-condensables.

20. Isolating chiller oil sumps

    a. helps collect all the oil.
    b. eliminates freon burn.
    c. reduces orifice plate damage.
    d. prevents refrigerant releases.

21. An azeotrope is?

    a. A leak testing instrument for HFCs only.
    b. A mixture of two or more refrigerants acting as one.
    c. A leak testing instrument for HFCs and HCFCs only.
    d. A leak testing instrument used with nitrogen.

22. Purge units should always

    a. be tested with air.
    b. be vented to mechanical room.
    c. be tested for leaks.
    d. be vented to the outside air.

23. A material used to line gloves to protect the body from refrigerants is

    a. latex rubber.   b. neoprene.
    c. butyl rubber.   d. cow hide.

24. Unlike HFC refrigerants, CFC and HCFC refrigerants contain

    a. hydrogen.   b. oxygen.
    c. hexaflouride.   d. chlorine.

25. 608 technicians must use recovery equipment as of this date.

    a. 5/14/93   b. 6/14/93
    c. 7/13/93   d. 8/12/93.

GROUP II SAMPLE TEST TYPE I TECHNICIAN

1. What is the smallest size container of R-22 that you may purchase by law?

   a. Any size available
   b. Anything over 5 lbs
   c. 20 lbs or more
   d. None of these

2. Replacing the condenser on a small window air conditioner is a

   a. minor repair.
   b. major repair.
   c. non recovery operation.
   d. requirement of leak repair.

3. The use of high and low side connections for recovery is used with

   a. elective equipment.
   b. type C equipment.
   c. self contained equipment.
   d. the push/pull method.

4. When transporting a disposable service jug to a job, it must

   a. be in a cardboard shipping container.
   b. Be secured in the vehicle.
   c. Be protected from the sun.
   d. (a) and (b).

5. Recovery is completed. Where may refrigerant be trapped?

   a. In the appliance accumulator.
   b. In the appliance drier.
   c. In the recovery hoses.
   d. None of these.

6. When would you evacuate your recovery machine with another recovery machine?

   a. To recover virgin refrigerant.
   b. Before recovering a different refrigerant.
   c. Before replacing filters.
   d. There is no need for it.

7. For these figures, 90/80/4 inHg, which is correct?

   a. Compressor operating/operating with leaks/non-condensables.
   b. Compressor inop/operating with leaks/minimum vacuum.
   c. Compressor with leaks / inoperative / non-condensables / minimum vacuum.
   d. Compressor operating / inoperative / minimum vacuum.

8. What must you keep records on?

   a. Refrigerants purchased, sold, recovered, reclaimed, recycled by quantity, type and service call.
   b. Recovery equipment registration and technician training and copy of certification.
   c. The number of small cans of R-12 purchased.
   d. (a) and (b)

9. What is the discharge or disassembly for deposit, dumping, placing, discarding or discharging on land or water?

   a. Pollution   b. Recovery
   c. Recycle   d. Disposal

10. What refrigerant(s) are compatible with polyolester oils?

    a. HCFC blends.
    b. azeotropic CFC blends.
    c. CFCs.
    d. HFC.

11. As a community service you prepare household appliances for disposal. One thing you must do is

    a. record who gave you the appliance.
    b. make a sworn statement to the disposer.
    c. recover 90/80/14inHg.
    d. (a), (b) and (c)

12. You have an empty disposable from liquid charging an appliance. What should you do with the disposable?

    a. Save it for use as an air tank.
    b. Use it for recovery of burned refrigerant.
    c. Recover any remaining refrigerant, then punch a hole in the vessel.
    d. Open the valve and put it in a dumpster.

13. A micron vacuum gauge is very useful for Type II and III technicians, but,

    a. it is just as useful for Type I technicians.
    b. it is unnecessary for Type I technicians.
    c. it can not be used with small appliances.
    d. it is only accurate for vacuums up to 15 inHg.

14. What does heating the compressor sump during vapor recovery help?

    a. Helps prelubricate the compressor bearings.
    b. Prevents a cold start.
    c. Helps lower the vapor pressure.
    d. Helps release refrigerant from the sump oil.

15. This method of recovery is best for leaving system oil.

    a. Type C   b. Vacuum
    c. Vapor   d. Liquid

16. You are purchasing a self contained recovery unit for use with multiple refrigerants. What should your unit have?

    a. Low loss fittings.
    b. Adjustable low and high pressure cutouts.
    c. Replaceable filters.
    d. Pump out cycle.

17. EPA requires you to have at least

    a. one piece of recovery equipment.
    b. one piece of certified recovery equipment.
    c. one piece of self contained recovery equipment.
    d. some recovery equipment.

18. What is the minimum recovery vessel capacity in pounds for Type I equipment?

    a. 6 lbs b. 5 lbs c. 8 lbs d. 10 lbs

19. When do you check your recovery equipment for leaks?

    a. Before each recovery.
    b. At least once a day.
    c. Once a month.
    d. After changing the filters.

20. One way to leak test an empty appliance is to

    a. 150 lbs of dry nitrogen and soap bubbles.
    b. 150 lbs of dry nitrogen and halide torch.
    c. 150 lbs of HCFC-22 and a sniffer.
    d. None of these.

21. One indicator of refrigerant overcharging is

    a. excessive compressor current.
    b. super cold evaporator air.
    c. increased superheat.
    d. higher than calculated condenser temperature.

22. A calibrated charging cylinder is connected to

    a. the center hose.
    b. the recovery units high side.
    c. the high side hose for liquid charging.
    d. the low side hose for vapor charging.

23. Liquid charging requires

    a. liquid into the suction side.
    b. liquid into the accumulator.
    c. the compressor to be off.
    d. a calibrated cylinder.

24. You are hired by a shop and receive training on their recovery unit using a window air conditioner, when you notice the recovery units vessel is white and has a 30 lb capacity. You should.

    a. Immediately stop and replace with a green vessel as you are recovering 5 lbs of R-22.
    b. Continue with the recovery, nothing is wrong.
    c. Replace it with a yellow vessel that indicates used refrigerant.
    d. Replace it with a vessel that has a grey body & yellow top.

25. You are leak testing a new condenser. What refrigerant should be used as a trace charge?

    a. CFC-12.
    b. CFC-12 or HCFC-22.
    c. HFC-134A and nitrogen.
    d. Several ounces of HCFC-22.

GROUP II TYPE II SAMPLE TEST

1. What are your recovery equipment requirements?

    a. Either system dependent or self contained.
    b. Self contained only.
    c. At least one piece of self contained.
    d. System dependent only.

2. How soon must leaks be repaired?

    a. Within 30 days of notice.
    b. Before EPA finds out.
    c. Never if an abatement plan is followed.
    d. Immediately.

3. The key to MVAC-like is

    a. non-road use.
    b. non-road use, mechanical vapor compression, open drive.
    c. agricultural use.
    d. uses R-12 for cooling.

4. Your new recovery unit has instructions that conflict with the EPA practices.

    a. Follow the factory instructions and notify EPA.
    b. Follow the EPA requirements.
    c. Recover using the factory instructions, then repeat using the EPA method.
    d. Do not use the equipment.

5. Your recovery unit has a higher low vacuum than required.
    a. That's okay, a stronger vacuum never hurt anything.
    b. Your recovery unit could burn up.
    c. The deeper vacuum could collapse a capillary tube.
    d. The deep vacuum may damage a pressure relief valve.

6. What is the annualized leak rate for retail food and cold storage appliances with 50 or more pounds of refrigerant?

    a. 1.5%.   b. 15%.
    c. 3.5%.   d. 35%.

7. Is it permissible to top off a system that is leaking?

    a. Yes, if it is to calculate the leak rate.
    b. If there is no leak repair requirement.
    c. Yes, if there is a retrofit or retirement plan in effect.
    d. All the above.

8. An appliance with 150 lbs. of HCFC-22 must be evacuated to what level using recovery equipment manufactured December 12, 1993?

    a. 15 inHg   b. 4 inHg
    c. 0 inHg   d. 25 inHg

9. Push/pull recovery requires

    a. any self contained recovery unit.

    b. a recovery vessel with liquid and vapor valves.

    c. the recovery output to a second vessel.

    d. all three.

10. De minimis protection only extends to the technician that follows what?

    a. Required practices and uses recovery equipment.

    b. Required practices, uses certified recovery equipment and complies with 82-36.

    c. 82-156, 82-154 and 82-158 and/or 82-36.

    d. None of these.

11. How is a ternary blend charged?

    a. Liquid or vapor is fine.

    b. Vapor only.

    c. Each gas separately.

    d. Liquid only.

12. EPA-established recovery levels must be met unless:

    a. Contamination of the refrigerant would result.

    b. There are leaks in the system.

    c. There is no "unless." Recovery levels must be met.

    d. Both (a) and (b).

13. When recovering refrigerants from large systems what precautions should you take?

    a. Before recovery, determine the capacity of your vessel.

    b. Check service valves.

    c. Look for areas that refrigerant could be trapped during recovery.

    d. All of these.

14. What must you do if the leak rate exceeds the EPA allowance?

    a. Notify EPA.

    b. Shut down the equipment.

    c. Evacuate the system.

    d. Notify the owner.

15. An appliance containing 350 lbs. of R-502 refrigerant has a leak. To what level must it be evacuated using equipment manufactured before November 14, 1993?

    a. 4 inHg for non leaking parts, 0 psig for leaking.

    b. 4 inHg leaking or not.

    c. 0 psig for leaking.

    d. 10 inHg.

16. An appliance with 350 lbs. of HCFC-22 has a leaking isolation valve. What is the minimum recovery requirement?

    a. 10 inHg.   b. 4 inHg.
    c. 0 psig.   d. 25 inHg.

17. Before recovering a different refrigerant with the same recovery unit you must

    a. replace the oil.

    b. replace the filters.

    c. change the vessel.

    d. (a) and (c)

18. MSDS stands for?

    a. Multi Stage Dehydration System.
    b. MonoSulpherDiesteroyl Synthetic.
    c. Material Safety Data Sheet.
    d. A sulphur dioxide based refrigerant.

19. You have certified recovery equipment certified by appendix B and/or C. Is it legal to use them to evacuate MVAC-like appliances?

    a. Yes, they meet 82-158 requirements.
    b. No, they do not meet the requirements of 82-36(a)
    c. Yes, if it is marked as per 82-158(h).
    d. Anything that is certified by EPA may be used.

20. Your recovery unit keeps shutting off during a push/pull recovery. What might be the cause?

    a. Non-condensables.
    b   Service valve in the port closed position.
    c. Clogged recovery unit inlet filter.
    d. Any of these.

21. For fastest recovery use

    a. a jet pump.

    b. short hoses.
    c. large diameter hoses.
    d. any of these.

22. What level of evacuation is required for MVAC-like appliances?

    a. To be below 102mm Hg.
    b. 4 in Hg.
    c. 15 in Hg.
    d. (a) and (b)

23. You added 105 lbs. of refrigerant to a cold storage warehouse appliance that has a capacity of 290 lbs. What is the leak rate?

    a. 33%.   b. 15%.
    c. 30%.   d. Over 35%.

24. System dependent equipment may always be used

    a. for R-12 recovery.
    b. for R-22 recovery.
    c. for any refrigerant recovery.
    d. for recovery of 15lbs or less.

25. You are recovering 100 lbs of refrigerant. You have two 50 lb vessels and two 25 lb vessels. Which ones should you use?

    a. Both 50 lb. and one 25 lb.
    b. Both 50 lb. vessels.
    c. Both 25 lb. and one 50 lb.
    d. All four.

GROUP II TYPE III SAMPLE TEST

1. Low pressure refrigerants are

    a. R-123, R-114, R-11.
    b. R-123, R-113, R-11.
    c. R-123, R-13, R-500.
    d. R-123, R-503.

2. A low pressure refrigerant has a boiling point

    a. above 10°C. b. below 10°C.
    c. above 15°C. d. below 15°C.

3. Push/pull is used because

    a. the compressed liquid pushes the vapor through the chiller tubes.
    b. the compressed vapor push causes thermal condensation of refrigerant in the vessel.
    c. the compressed liquid enters(push) the liquid valve, boils, and the pressure drop pulls the liquid in.
    d. it is fast and cost effective.

4. The required level of pre and post evacuation required for low pressure appliances is

    a. 25 inHg/25 inHg.
    b. 25 mm Hg absolute/25 inHg.
    c. 25 mm Hg/25 mm Hg.
    d. 25 inHg/25 mm Hg absolute.

5. A low pressure appliance is leaking a ternary blend of MP-39. Why must you evacuate and recharge the system AFTER the leak is fixed?

    a. A fresh liquid recharge is the only way to ensure proper blend ratio and thermal glide.
    b. A fresh vapor recharge is the only way to ensure proper blend ratio and thermal glide.
    c. If you can fix the leak without evacuation then it is OK to top off with vapor.
    d. If you can fix the leak without evacuation then it is OK to top off with liquid.

6. The most frequent cause of job site accidental death of refrigerant technicians is

    a. electrocution.   b. poisoning.
    c. asphyxiation.   d. falls.

7. Where would SCBA be used?

    a. In the shop.
    b. In a work room.
    c. In a basement.
    d. Anywhere that refrigerants could accumulate.

8. The easiest way to crack chiller tubes is to

    a. leave heaters on with no water in system.
    b. liquid charge from a deep vacuum.
    c. use ultrasonic tester on empty tubes.
    d. use ultrasonic tester on full tubes.

9   CFC refrigerants will not be manufactured after

    a.   November 14, 1994.
    b.   December 31, 1994.
    c.   December 31, 1995.
    d.   November 30, 1994.

10.   You find a leak rate over 40%. Who is responsible for repair of the appliance?

    a.   You found it, you fix it.
    b.   EPA regulations require you to fix it.
    c.   The owner is responsible.
    d.   No requirement if less than 50 lbs. charge.

11.   Push/pull recovery

    a.   active recovery.
    b.   passive recovery.
    c.   inert recovery.
    d.   None of the above.

12.   Why is leak detection so important?

    a.   Due to cost of refrigerants.
    b.   It reduces purging releases.
    c.   It maintains system efficiency.
    d.   All the above.

13.   You are ready to charge a chiller. What should be running?

    a.   The cooling fan.
    b.   The purge unit.
    c.   Circulation pump.
    d.   Nothing, just charge it.

14.   A chiller has multiple leaks. Specified recovery levels cannot be met due to pulling air into the system and recovery unit. What is your minimum action?

    a.   Pack up and go home.
    b.   Pressurize system to 0 psig.
    c.   Evacuate to 0 psig.
    d.   Evacuate to 29 inHg.

15.   Where should the purge drum non-condensables go?

    a.   To the outside.
    b.   To a recovery vessel.
    c.   Vent to the equipment room.
    d.   Store in a recovery tank for disposal.

16.   What is connected to the evaporator of all low pressure equipment?

    a.   Non-condensable line.
    b.   Condensable line.
    c.   Rupture disk.
    d.   Oxygen depletion monitor.

17.   The term 25mm Hg. absolute means?

    a.   A vacuum of 27 inHg.
    b.   The same as 25 inHg.
    c.   29 in Hg gauge.
    d.   30 in Hg gauge.

18.   The best way to keep costs low and maintain system efficiencies is to

    a.   repair leaks.
    b.   recycle refrigerant.
    c.   track refrigerant use.
    d.   All the above.

19. Any time appliance pressure is being changed for leak testing, what is most important?

    a. Finding the leak.
    b. The type of leak detection equipment to use.
    c. Do not exceed 10 psig system pressure.
    d. The refrigerant you are testing for.

20. Where is the rupture disc located on low pressure appliances?

    a. Next to the purge drum pull-off line.
    b. At the compressor discharge port.
    c. On the condenser input side.
    d. On the evaporator input side.

21. Will a rupture disc fail if a chiller is charged with 200 lbs. of R-11?
a. Yes.   b. No.

22. At what pressure will a rupture disc open?

a. 15 psig.   b. 10 psig.
c. 20 psig.   d. 190 lbs.

23. PPE should be

    a. butyl rubber lined gloves.
    b. full face shield.
    c. butyl rubber body apron.
    d. all of these.

24. Where does the purge drum connect?

    a. Evaporator output.
    b. On the accumulator.
    c. On top of the condenser.
    d. Any of these.

25. Comfort cooling chillers with 50 lbs. of charge or more must have leaks repaired if the leak exceeds

a. 35%.   b. 15%.
c. any amount.   d. 30%.

SEE PAGE 124 FOR ANSWER KEY

GROUP I ANSWERS

1.	A	6.	A	11.	D	16.	C	21.	B
2.	D	7.	A	12.	C	17.	D	22.	D
3.	B	8.	D	13.	B	18.	C	23.	C
4.	A	9.	AorC	14.	C	19.	C	24.	D
5.	C	10.	D	15.	D	20.	D	25.	D

GROUP II TYPE I ANSWERS

1.	C	6.	B	11.	B	16.	D	21.	A
2.	B	7.	D	12.	C	17.	B	22.	A
3.	B	8.	D	13.	A	18.	C	23.	C
4.	D	9.	D	14.	D	19.	B	24.	D
5.	C	10.	D	15.	C	20.	A	25.	D

GROUP II TYPE II ANSWERS

1.	C	6.	D	11.	D	16.	C	21.	C
2.	C	7.	D	12.	D	17.	D	22.	A
3.	B	8.	C	13.	D	18.	C	23.	D
4.	B	9.	A	14.	D	19.	B	24.	D
5.	B	10.	C	15.	A	20.	D	25.	A

GROUP II TYPE III ANSWERS

1.	B	6.	C	11.	A	16.	C	21.	B
2.	A	7.	D	12.	D	17.	C	22.	A
3.	D	8.	B	13.	C	18.	D	23.	D
4.	C	9.	C	14.	B	19.	C	24.	C
5.	A	10.	C&D	15.	A	20.	D	25.	B

APPENDIX IV

FEDERAL REGISTER CFR 40 Part 82 (Section 608)

The following Code of Federal Regulations (CFR) is printed in a larger readable type than the published versions that can be obtained from the Government Printing Office. This Appendix excludes the 40 CFR Part 82 summary and Appendices. The majority of the summary material and Appendices data is covered throughout this text. Official copies of the summary and complete regulations can be obtained from the Government Printing Office (GPO) at 202-512-1530 or by calling the EPA Hotline at 1-800-296-1996. If you have a personal computer with a modem and communications program, you can download an unofficial copy of 40 CFR Part 82 (file 608RULE.ZIP) from the EPA's TTN Clean Air Act electronic bulletin board. The access number is 919-541-5742.

This chapter was downloaded electronically from EPA's bulletin board. The EPA cautions that, "the electronic dissemination of EPA files is an experimental project of EPA and the GPO. While every effort has been made to ensure that these electronic files accurately reflect the contents of the CFR, this **IS NOT AN OFFICIAL VERSION** and it **SHOULD NOT BE RELIED UPON** for purposes of legal enforcement or citation. In the event that the material in the electronic files varies from the text of the CFR, the official version as published by the Administrative Committee of the Federal Register (ACFR) controls." Obtain official copies for your shop from the resources listed above.

This CFR copy was published prior to several of the EPA's final decisions. The final rule dates are interspersed throughout the study material of this book and specifically in Chapter Three. The sample test questions will also specify the final rule dates that were available at the time this book was printed.

125

The author and publisher make no warranties, either expressed or implied, with respect to the information contained herein. This CFR was provided by the EPA bulletin board as mentioned above. The author and publisher shall not be liable for any incidental or consequential damages in connection with, or arising out of, the use of material in this book.

CFR 40 Part 82 (Section 608)

For the reasons set out in the Preamble, EPA is amending 40 CFR part 82 as follows:

PART 82--PROTECTION OF STRATOSPHERIC OZONE

1. Authority: The authority citation for part 82 is as follows:

Authority: 42 U.S.C. §7414, 7601, §7671-7671q.

2. Part 82 is amended by adding Subpart F to read as follows:

Subpart F--Recycling and Emissions Reduction

Sec.
82.150 Purpose and Scope
82.152 Definitions
82.154 Prohibitions
82.156 Required Practices
82.158 Standards for Recycling   and Recovery Equipment
82.160 Approved Equipment   Testing Organizations
82.161 Technician   Certification
82.162 Certification by   Owners of Recovery   and Recycling   Equipment
82.164 Reclaimer Certification
82.166 Reporting and Record-   keeping Requirements

Appendix A to Subpart F-- ARI Standard 700-1988, Specifications for Fluorocarbon Refrigerants
Appendix B to Subpart F-- ARI Standard 740-1993, Performance of Refrigerant Recovery, Recycling and/or Reclaim Equipment
Appendix C to Subpart F-- Method for Testing Recovery Devices for Use with Small Appliances
Appendix D to Subpart F-- Standards for Becoming a Certifying Program for Technicians

Subpart F--Recycling and Emissions Reduction
§82.150 Purpose and Scope

(a) The purpose of this Subpart is to reduce emissions of class I and class II refrigerants to the lowest achievable level during the service, maintenance, repair, and disposal of appliances in accordance with section 608 of the Clean Air Act.

(b) This Subpart applies to any person servicing, maintaining, or repairing appliances except for motor vehicle air conditioners. This subpart also applies to persons disposing of appliances, including motor vehicle air conditioners. In addition, this subpart applies to refrigerant reclaimers, appliance owners, and manufacturers of appliances and recycling and recovery equipment.

§82.152 Definitions.

(a) Appliance means any device which contains and uses a class I or class II substance as a refrigerant and which is used for household or commercial purposes, including any air conditioner, refrigerator, chiller, or freezer.

(b) Approved Equipment Testing Organization means any organization which has applied for and received approval from the Administrator pursuant to §82.160.

(c) Certified Refrigerant Recovery or Recycling Equipment means equipment certified by an approved equipment testing organization to meet the standards in §82.158(b) or (d), equipment certified pursuant to §82.36(a), or equipment manufactured before November 15, 1993 that meets the standards in §82.158(c), (e), or (g).

(d) Commercial Refrigeration means, for the purposes of § 82.156(i), the refrigeration appliances utilized in the retail food and cold storage warehouse sectors. Retail food includes the refrigeration equipment found in supermarkets, convenience stores, restaurants and other food service establishments. Cold storage includes the equipment used to store meat, produce, dairy products, and other perishable goods. All of the equipment contains large refrigerant charges, typically over 75 pounds.

(e) Disposal means the process leading to and including
(1) the discharge, deposit, dumping or placing of any discarded appliance into or on any land or water,
(2) the disassembly of any appliance for discharge, deposit, dumping or placing of its discarded component parts into or on any land or water, or
(3) the disassembly of any appliance for reuse of its component parts.

(f) High-Pressure Appliance means an appliance that uses a refrigerant with a boiling point between -50 and 10 degrees Centigrade at atmospheric pressure (29.9 inches of mercury). This

definition includes but is not limited to appliances using refrigerants-12, -22, -114, -500, or -502.

(g) Industrial Process Refrigeration means, for the purposes of § 82.156(i), complex customized appliances used in the chemical, pharmaceutical, petrochemical and manufacturing industries. This sector also includes industrial ice machines and ice rinks.

(h) Low-loss fitting means any device that is intended to establish a connection between hoses, appliances, or recovery or recycling machines and that is designed to close automatically or to be closed manually when disconnected, minimizing the release of refrigerant from hoses, appliances, and recovery or recycling machines.

(i) Low-Pressure Appliance means an appliance that uses a refrigerant with a boiling point above 10 degrees Centigrade at atmospheric pressure (29.9 inches of mercury). This definition includes but is not limited to equipment utilizing refrigerants-11, -113, and -123.

(j) Major maintenance, service, or repair means any maintenance, service, or repair involving the removal of any or all of the following

appliance components: compressor, condenser, evaporator, or auxiliary heat exchanger coil.

(k) Motor Vehicle Air Conditioner (MVAC) means any appliance that is a motor vehicle air conditioner as defined in 40 CFR Part 82 Subpart B.

(l) MVAC-like Appliance means mechanical vapor compression, open-drive compressor appliances used to cool the driver's or passenger's compartment of an non-road motor vehicle. This includes the air-conditioning equipment found on agricultural or construction vehicles. This definition is not intended to cover appliances using HCFC-22 refrigerant.

(m) Normally Containing a quantity of refrigerant means containing the quantity of refrigerant within the appliance or appliance component when the appliance is operating with a full charge of refrigerant.

(n) Opening an appliance means any service, maintenance, or repair on an appliance that could be reasonably expected to release refrigerant from the appliance to the atmosphere unless the refrigerant were previously recovered from the appliance.

(o) Person means any

individual or legal entity, including an individual, corporation, partnership, association, state, municipality, political subdivision of a state, Indian tribe, and any agency, department, or instrumentality of the United States, and any officer, agent, or employee thereof.

(p) Process stub means a length of tubing that provides access to the refrigerant inside a small appliance or room air conditioner and that can be resealed at the conclusion of repair or service.

(q) Reclaim refrigerant means to reprocess refrigerant to at least the purity specified in the ARI Standard 700-1988, Specifications for Fluorocarbon Refrigerants (Appendix A to 40 CFR Part 82 Subpart F) and to verify this purity using the analytical methodology prescribed in the ARI Standard 700-1988. In general, reclamation involves the use of processes or procedures available only at a reprocessing or manufacturing facility.

(r) Recover refrigerant means to remove refrigerant in any condition from an appliance without necessarily testing or processing it in any way.

(s) Recovery Efficiency means the percentage of refrigerant in an appliance that is recovered by a piece of recycling or recovery equipment.

(t) Recycle refrigerant means to extract refrigerant from an appliance and clean refrigerant for reuse without meeting all of the requirements for reclamation. In general, recycled refrigerant is refrigerant that is cleaned using oil separation and single or multiple passes through devices, such as replaceable core filter-driers, which reduce moisture, acidity, and particulate matter. These procedures are usually implemented at the field job site.

(u) Self-Contained Recovery Equipment means refrigerant recovery or recycling equipment that is capable of removing the refrigerant from an appliance without the assistance of components contained in the appliance.

(v) Small appliance means any of the following products that are fully manufactured, charged, and hermetically sealed in a factory with five (5) pounds or less of refrigerant: refrigerators and freezers designed for home use, room air conditioners (including window air conditioners and packaged terminal air conditioners), packaged terminal heat pumps, dehumidifiers, under-the-counter ice makers, vending machines, and drinking water coolers.

(w) System-Dependent Recovery Equipment means refrigerant recovery equipment that requires the assistance of components contained in an appliance to remove the refrigerant from the appliance.

(x) Technician means any person who performs maintenance, service, or repair that could reasonably be expected to release class I or class II substances from appliances into the atmosphere, including but not limited to installers, contractor employees, in-house service personnel, and in some cases, owners. Technician also means any person disposing of appliances except for small appliances.

(y) Very High-Pressure Appliance means an appliance that uses a refrigerant with a boiling point below -50 degrees Centigrade at atmospheric pressure (29.9 inches of mercury). This definition includes but is not limited to equipment utilizing refrigerants-13 and -503.

§82.154 Prohibitions
(a) Effective June 14, 1993, no person maintaining, servicing, repairing, or disposing of appliances may knowingly vent or otherwise release into the environment any class I or class II substance used as refrigerant in such equipment. De

minimis releases associated with good faith attempts to recycle or recover refrigerants are not subject to this prohibition. Releases shall be considered de minimis if they occur when:

(1) the required practices set forth in §82.156 are observed and recovery or recycling machines that meet the requirements set forth in §82.158 are used, or

(2) the requirements set forth in 40 CFR Part 82 Subpart B are observed.

The knowing release of refrigerant subsequent to its recovery from an appliance shall be considered a violation of this prohibition.

(b) Effective July 13, 1993, no person may open appliances except MVACs for maintenance, service, or repair, and no person may dispose of appliances except for small appliances, MVACs, and MVAC-like appliances,

(1) Without observing the required practices set forth in §82.156 and

(2) Without using equipment that is certified for that type of appliance pursuant to §82.158.

(c) Effective November 15, 1993, no person may manufacture or import recycling or recovery equipment for use during the maintenance, service, or repair of appliances except MVACs, and no person may manufacture or import recycling or recovery equipment for use during the disposal of appliances except small appliances, MVACs, and MVAC-like appliances, unless the equipment is certified pursuant to §82.158(b), (d), or (f), as applicable.

(d) Effective June 14, 1993, no person shall alter the design of certified refrigerant recycling or recovery equipment in a way that would affect the equipment's ability to meet the certification standards set forth in §82.158 without resubmitting the altered design for certification testing. Until it is tested and shown to meet the certification standards set forth in §82.158, equipment so altered will be considered uncertified for the purposes of §82.158.

(e) Effective August 12, 1993, no person may open appliances except MVACs for maintenance, service, or repair, and no person may dispose of appliances except for small appliances, MVACs, and MVAC-like appliances, unless such person has certified to the Administrator pursuant to §82.162 that such person has acquired certified recovery or recycling equipment and is complying with the applicable requirements of this rule.

(f) Effective August 12, 1993, no person may recover refrigerant from small appliances, MVACs, and MVAC-like appliances for purposes of disposal of these appliances unless such person has certified to the Administrator pursuant to §82.162 that such person has acquired recovery equipment that meets the standards set forth in §82.158(l) and/or (m), as applicable, and that such person is complying with the applicable requirements of this rule.

(g) Effective August 12, 1993 until May 15 1995, no person may sell or offer for sale for use as a refrigerant any class I or class II substance consisting wholly or in part of used refrigerant unless the class I or class II substance has been reclaimed as defined at §82.152 (q).

(h) Effective August 12, 1993 until May 15, 1995, no person may sell or offer for sale for use as a refrigerant any class I or class II substance consisting wholly or in part of used refrigerant unless the refrigerant has been reclaimed by a person who has been certified as a reclaimer pursuant to §82.164.

(i) Effective August 12, 1993, no person reclaiming refrigerant may release more than 1.5% of the refrigerant received by them.

(j) Effective November

15, 1993, no person may sell or distribute, or offer for sale or distribution, any appliances, except small appliances, unless such equipment is equipped with a servicing aperture to facilitate the removal of refrigerant at servicing and disposal.

(k) Effective November 15, 1993, no person may sell or distribute, or offer for sale or distribution any small appliance unless such equipment is equipped with a process stub to facilitate the removal of refrigerant at servicing and disposal.

(l) Effective November 14, 1993 no person may open an appliance except for an MVAC and no person may dispose of an appliance except for a small appliance, MVAC, or MVAC-like appliance, unless such person has been certified as a technician for that type of appliance pursuant to §82.161.

(m) No technician training or testing program may issue certificates pursuant to §82.161 unless the program complies with all of the standards of §82.161 and Appendix D, and has been granted approval.

(n) Effective November 14, 1994 no person may sell or distribute, or offer for sale or distribution, any class I or class II substance for use as a refrigerant to any person

unless:

(1) the buyer has been certified as a Type I, Type II, Type III, or Universal technician pursuant to §82.161,

(2) the buyer has been certified pursuant to 40 CFR Part 82 Subpart B,

(3) the refrigerant is sold only for eventual resale to certified technicians or to appliance manufacturers (e.g., sold by a manufacturer to a wholesaler, sold by a technician to a reclaimer),

(4) the refrigerant is sold to an appliance manufacturer,

(5) the refrigerant is contained in an appliance, or

(6) the refrigerant is charged into an appliance by a certified technician during maintenance, service, or repair.

(o) It is a violation of these regulations to accept a signed statement pursuant to §82.156 (f)(2) if the person knew or had reason to know that such a signed statement is false.

§82.156 Required Practices

(a) Effective July 13, 1993, all persons opening appliances except for MVACs for maintenance, service, or repair must evacuate the refrigerant in either the entire unit or the part to be serviced (if the latter can be isolated) to a system receiver or a recovery or recycling machine certified pursuant to §82.158. All persons disposing of

appliances except for small appliances, MVACs, and MVAC-like appliances must evacuate the refrigerant in the entire unit to a recovery or recycling machine certified pursuant to §82.158.

(1) Persons opening appliances except for small appliances, MVACs, and MVAC-like appliances for maintenance, service, or repair must evacuate to the levels in Table 1 before opening the appliance, unless

(i) evacuation of the appliance to the atmosphere is not to be performed after completion of the maintenance, service, or repair, and the maintenance, service, or repair is not major as defined at §82.152(j), or

(ii) due to leaks in the appliance, evacuation to the levels in Table 1 is not attainable, or would substantially contaminate the refrigerant being recovered. In any of these cases, the requirements of §82.156(a)(2) must be followed.

(2)(i) If evacuation of the appliance to the atmosphere is not to be performed after completion of the maintenance, service, or repair, and if the maintenance, service, or repair is not major as defined at §82.152(j), the appliance must:

(A) be evacuated to a pressure no higher than 0 psig before it is opened if it is a high- or very high-pressure appliance, or

(B) be pressurized to 0

psig before it is opened if it is a low-pressure appliance, without using methods, e.g., nitrogen, that require subsequent purging.

(ii) If, due to leaks in the appliance, evacuation to the levels in Table 1 is not attainable, or would substantially contaminate the refrigerant being recovered,

persons opening the appliance must:

(A) isolate leaking from non-leaking components wherever possible,

(B) evacuate non-leaking components to be opened to the levels specified in Table 1, and

(C) evacuate leaking components to be opened to

the lowest level that can be attained without substantially contaminating the refrigerant. In no case shall this level exceed 0 psig.

(3) Persons disposing of appliances except for small appliances, MVACs, and MVAC-like appliances, must evacuate to the levels in Table 1.

Table 1

Required Levels of Evacuation for Appliances Except for Small Appliances, MVACs, & MVAC-like Appliances

Inches of Hg Vacuum (Relative to Standard Atmospheric Pressure of 29.9 Inches Hg)

Type of Appliance	Using Recovery or Recycling Equipment Manufactured or Imported Before [6 months after publication of final rule]	Using Recovery or Recycling Equipment Manufactured or Imported On or After [6 months after publication of final rule]
HCFC-22 appliance, or isolated com-ponent of such appliance, normally containing less than 200 pounds of refrigerant	0	0
HCFC-22 appliance, or isolated component of such appliance, normally containing 200 pounds or more of refrigerant	4	10
Other high-pressure appliance, or isolated component of such appliance, normally containing less than 200 pounds of refrigerant	4	10
Other high-pressure appliance, or isolated component of such appliance, normally containing 200 pounds or more of refrigerant	4	15
Very High-Pressure Appliance	0	0
Low-Pressure Appliance	25	25 mm Hg absolute

(4) Persons opening small appliances for maintenance, service, or repair must

(i) when using recycling and recovery equipment manufactured before [six months after publication of the final rule], recover 80% of the refrigerant in the small

appliance, or

(ii) when using recycling or recovery equipment manufactured on or after [six months after publication of the final rule], recover 90% of the refrigerant in the appliance when the compressor in the appliance is operating, or 80%

of the refrigerant in the appliance when the compressor in the appliance is not operating, or

(iii) evacuate the small appliance to four inches of mercury vacuum.

(5) Persons opening MVAC-like appliances for

maintenance, service, or repair may do so only while properly using, as defined at 40 CFR §82.32(e), recycling or recovery equipment certified pursuant to §82.158(f) or (g), as applicable.

(b) Effective July 13, 1993, all persons opening appliances except for small appliances and MVACs for maintenance, service, or repair and all persons disposing of appliances except for small appliances must have at least one piece of certified, self-contained recovery equipment available at their place of business.

(c) System-dependent equipment shall not be used with appliances normally containing more than 15 pounds of refrigerant.

(d) All recovery or recycling equipment shall be used in accordance with the manufacturer's directions unless such directions conflict with the requirements of this rule.

(e) Refrigerant may be returned to the appliance from which it is recovered or to another appliance owned by the same person without being recycled or reclaimed, unless the appliance is an MVAC-like appliance.

(f) Effective July 13, 1993, persons who take the final step in the disposal process (including but not limited to scrap recyclers and landfill operators) of a small appliance, room air conditioning, MVACs, or MVAC-like appliances must either

(1) recover any remaining refrigerant from the appliance in accordance with paragraph (g) or (h) below, as applicable, or

(2) verify that the refrigerant has been evacuated from the appliance or shipment of appliances previously. Such verification must include a signed statement from the person from whom the appliance or shipment of appliances is obtained that all refrigerant that had not leaked previously has been recovered from the appliance or shipment of appliances in accordance with paragraph (g) or (h) below, as applicable. This statement must include the name and address of the person who recovered the refrigerant and the date the refrigerant was recovered or a contract that refrigerant will be removed prior to delivery.

(3) Persons complying with (f)(2) above must notify suppliers of appliances that refrigerant must be properly removed before delivery of the items to the facility. The form of this notification may be warning signs, letters to suppliers, or other equivalent means.

(g) All persons recovering refrigerant from MVACs and MVAC-like appliances for purposes of disposal of these appliances must reduce the system pressure to or below 102 mm of mercury vacuum, using equipment that meets the standards set forth in §82.158 (l).

(h) All persons recovering the refrigerant from small appliances for purposes of disposal of these appliances must either

(i) recover 90% of the refrigerant in the appliance when the compressor in the appliance is operating, or 80% of the refrigerant in the appliance when the compressor in the appliance is not operating, or

(ii) evacuate the small appliance to four inches of mercury vacuum.

(i) (1) Owners of commercial refrigeration and industrial process refrigeration equipment must have all leaks repaired if the equipment is leaking at a rate such that the loss of refrigerant will exceed 35 percent of the total charge during a 12 month period, except as described in (i)(3) below.

(2) Owners of appliances normally containing more than 50 pounds of refrigerant and not covered by (i)(1) above must have all leaks repaired if the appliance is leaking at a rate such that the loss of refrigerant will exceed 15 % of the total charge during a

12-month period, except as described in (i)(3) below.

(3) Owners are not required to repair the leaks defined in (i)(1) and (2) above if, within 30 days, they develop a one-year retrofit or retirement plan for the leaking equipment. This plan (or a legible copy) must be kept at the site of the equipment. The original must be made available for EPA inspection on request. The plan must be dated and all work under the plan must be completed within one year of plan's date.

(4) Owners must repair leaks pursuant to (i)(1) and (2) above within 30 days of discovery or within 30 days of when the leak(s) should have been discovered, if the owners intentionally shielded themselves from information which would have revealed a leak.

[Approved by the Office of Management and Budget under the control number 2060-0256]

§82.158 Standards for Recycling and Recovery Equipment

(a) Effective November 15, 1993, all manufacturers and importers of recycling and recovery equipment intended for use during the maintenance, service, or repair of appliances except MVACs and MVAC-like appliances or during the disposal of appliances except small appliances, MVACs, and MVAC-like appliances, shall have had such equipment certified by an approved equipment testing organization to meet the applicable requirements in (b) or (d) below. All manufacturers and importers of recycling and recovery equipment intended for use during the maintenance, service, or repair of MVAC-like appliances shall have had such equipment certified pursuant to §82.36(a).

(b) Equipment manufactured or imported on or after [6 months after publication of the final rule] for use during the maintenance, service, or repair of appliances except small appliances, MVACs, and MVAC-like appliances or during the disposal of appliances except small appliances, MVACs, and MVAC-like appliances must be certified by an approved equipment testing organization to meet the following requirements:

(1) In order to be certified, the equipment must be capable of achieving the level of evacuation specified in Table 2 below under the conditions of the ARI Standard 740-1993, Performance of Refrigerant Recovery, Recycling and/or Reclaim Equipment (ARI 740-1993)(Appendix B):

Table 2
Levels of Evacuation which Must Be Achieved by
Recovery or Recycling Equipment Intended for Use With Appliances

(Except for Small Appliances, MVACs, and MVAC-like Appliances)
Manufactured On or After [6 months after publication of the final rule]

Type of Appliance with which Recovery or Recycling Machine is Intended to be Used	Inches of Hg Vacuum
HCFC-22 appliances, or isolated component of such appliances, normally containing less than 200 pounds of refrigerant	0
HCFC-22 appliances, or isolated components of such appliances, normally containing 200 pounds or more of refrigerant	10
Very high-pressure appliances	0
Other high-pressure appliances, or isolated component of such appliances, normally containing less than 200 pounds of refrigerant	10
Other high-pressure appliances, or isolated component of such appliances, normally containing 200 pounds or more of refrigerant	15
Low-pressure appliances	25 mm Hg absolute

The vacuums specified in inches of Hg vacuum must be achieved relative to an atmospheric pressure of 29.9 inches of Hg absolute.

(2) Recovery or recycling equipment whose recovery efficiency cannot be tested according to the procedures in ARI 740-1993 may be certified if an approved third-party testing organization adopts and performs a test that demonstrates, to the satisfaction of the Administrator, that the recovery efficiency of that equipment is equal to or better than that of equipment that

(i) is intended for use with the same type of appliance and   (ii) achieves the level of evacuation in Table 2.

(3) The equipment must meet the minimum requirements for ARI certification under ARI 740-1993.

(4) If the equipment is equipped with a noncondensables purge device,

(i) the equipment must not release more than five percent of the quantity of refrigerant being recycled through noncondensables purging under the conditions of

ARI 740-1993, and

(ii) effective [two years after publication of the final rule], the equipment must not release more than three percent of the quantity of refrigerant being recycled through noncondensables purging under the conditions of ARI 740-1993.

(5) The equipment must be equipped with low-loss fittings on all hoses.

(6) The equipment must have its liquid recovery rate and its vapor recovery rate measured under the conditions of ARI 740-1993.

(c) Equipment manufactured or imported before [6 months after publication of the final rule] for use during the maintenance, service, or repair of appliances except small appliances, MVACs, and MVAC-like appliances or during the disposal of appliances except small appliances, MVACs, and MVAC-like appliances will be considered certified if it is capable of achieving the level of evacuation specified in Table 3 below when tested using a properly calibrated pressure gauge:

Table 3

Levels of Evacuation which
Must Be Achieved by Recovery or Recycling Machines Intended for
Use With Appliances
(Except for Small Appliances, MVACs, and MVAC-like Appliances)
Manufactured Before [6 months after publication of the final rule]

Type of Air-conditioning or Refrigeration Equipment with which Recovery or Recycling Machine is Intended to be Used	Inches of Vacuum (Relative to Standard Atmospheric Pressure of 29.9 Inches Hg)
HCFC-22 equipment, or isolated component of such equipment, normally containing less than 200 pounds of refrigerant	0
HCFC-22 equipment, or isolated component of such equipment, normally containing 200 pounds or more of refrigerant	4
Very high-pressure equipment	0
Other high-pressure equipment, or isolated component of such equipment, normally containing less than 200 pounds of refrigerant	4
Other high-pressure equipment, or isolated component of such equipment, normally containing 200 pounds or more of refrigerant	4
Low-pressure equipment	25

(d) Equipment manufactured or imported on or after [6 months after publication of the final rule] for use during the maintenance, service, or repair of small appliances must be certified by an approved equipment testing organization to be capable of either

(i) recovering 90% of the refrigerant in the test stand when the compressor of the test stand is operating and 80% of the refrigerant when the compressor of the test stand is not operating when used in accordance with the manufacturer's instructions under the conditions of Appendix C, Method for Testing Recovery Devices for Use with Small Appliances, or

(ii) achieving a four-inch vacuum under the conditions of Appendix B, ARI 740-1993.

(e) Equipment manufactured or imported before November 15, 1993 for use with small appliances will be considered certified if it is capable of either

(i) recovering 80% of the refrigerant in the system, whether or not the compressor of the test stand is operating, when used in accordance with the manufacturer's instructions under the conditions of Appendix C, Method for Testing Recovery Devices for Use with Small Appliances, or

(ii) achieving a four-inch vacuum when tested using a properly calibrated pressure gauge.

(f) Equipment manufactured or imported on or after November 15, 1993 for use during the maintenance, service, or repair of MVAC-like appliances must be certified in accordance with 40 CFR §82.36(a).

(g) Equipment manufactured or imported before [6 months after publication of the final rule] for use during the maintenance, service, or repair of MVAC-like appliances must be capable of reducing the system pressure to 102 mm of mercury vacuum under the conditions of the SAE Standard, SAE J1990 (Appendix A to 40 CFR Part 82 Subpart B).

(h) Manufacturers and importers of equipment certified under paragraphs (b) and (d) of this section must place a label on each piece of equipment stating the following:
THIS EQUIPMENT HAS BEEN CERTIFIED BY [APPROVED EQUIPMENT TESTING ORGANIZATION] TO MEET EPA'S MINIMUM REQUIREMENTS FOR RECYCLING OR RECOVERY EQUIPMENT INTENDED FOR USE WITH [APPROPRIATE CATEGORY OF APPLIANCE].
The label shall also show the date of manufacture and the serial number (if applicable) of the equipment. The label shall be affixed in a readily visible or accessible location, be made of a material expected to last the lifetime of the equipment, present required information in a manner so that it is likely to remain legible for the lifetime of the equipment, and be affixed in such a manner that it cannot be removed from the equipment without damage to the label.

(i) The Administrator will maintain a list of equipment certified pursuant to paragraphs (b), (d), and (f) above by manufacturer and model. Persons interested in obtaining a copy of the list should send written inquiries to the address in paragraph 82.160 (a) of this section.

(j) Manufacturers or importers of recycling or recovery equipment intended for use during the maintenance, service, or repair of appliances except MVACs or MVAC-like appliances or during the disposal of appliances except small appliances, MVACs, and MVAC-like appliances must periodically have approved equipment testing organizations conduct either

(1) retests of certified recycling or recovery equipment or

(2) inspections of recycling or recovery equipment at manufacturing facilities to ensure that each equipment model line that has been certified under this section continues to meet the certification criteria.

Such retests or inspections must be conducted at least once every three years after the equipment is first certified.

(k) An equipment model line that has been certified under this section may have its certification revoked if it is subsequently determined to fail to meet the certification criteria. In such cases, the Administrator or her or his designated representative shall give notice to the manufacturer or importer setting forth the basis for her or his determination.

(l) Equipment used to evacuate refrigerant from MVACs and MVAC-like appliances before they are disposed of must be capable of reducing the system pressure to 102 mm of mercury vacuum under the conditions of the SAE Standard, SAE J1990 (Appendix A to 40 CFR Part 82 Subpart B).

(m) Equipment used to evacuate refrigerant from small appliances before they are disposed of must be capable of either

(i) removing 90% of the refrigerant when the compressor of the small appliance is operating and 80% of the refrigerant when the compressor of the small appliance is not operating, when used in accordance with the manufacturer's instructions under the conditions of Appendix C, Method for Testing Recovery Devices for Use with Small Appliances, or

(ii) evacuating the small appliance to four inches of vacuum when tested using a properly calibrated pressure gauge.

§82.160 Approved Equipment Testing Organizations

(a) Any equipment testing organization may apply for approval by the Administrator to certify equipment pursuant to the standards in §82.158 and appendices B or C. The application shall be sent to:
§608 Recycling Program Manager
Stratospheric Protection Division
6205-J, U.S. Environmental Protection Agency
401 M Street, SW.
Washington, DC 20460

(b) Applications for approval must include written information verifying the following:

(1) The list of equipment present at the organization that will be used for equipment testing.

(2) Expertise in equipment testing and the technical experience of the organization's personnel.

(3) Thorough knowledge of the standards as they appear in §82.158 and Appendices B and/or C (as applicable) of this subpart.

(4) The organization must describe its program for verifying the performance of certified recycling and recovery equipment manufactured over the long term, specifying whether retests of equipment or inspections of equipment at manufacturing facilities will be used.

(5) The organization must have no conflict of interest and receive no direct or indirect financial benefit from the outcome of certification testing.

(6) The organization must agree to allow the Administrator access to records and personnel to verify the information contained in the application.

(c) Organizations may not certify equipment prior to receiving approval from EPA. If approval is denied under this section, the Administrator or her or his designated representative shall give written notice to the organization setting forth the basis for her or his determination.

(d) If at any time an approved testing organization is found to be conducting certification tests for the purposes of this subpart in a manner not consistent with the representations made in its application for approval under this section, the Administrator reserves the right to revoke approval. In such cases, the Administrator or her or his designated representative shall give notice to the organization setting forth the basis for her or his determination.

(e) Testing organizations seeking approval of an equipment certification program may also seek approval to certify equipment tested previously under the program. Interested organizations may submit to the Administrator at the address in §82.160(a) verification that the program met all of the standards in §82.160(b) and that equipment to be certified was tested to and met the applicable standards in §82.158(b) or (d). Upon EPA approval, the previously tested equipment may be certified without being retested (except insofar as such retesting is part of the testing organization's program for verifying the performance of equipment manufactured over the long term, pursuant to §82.160(b)(4)).

[Approved by the Office of Management and Budget under the control number 2060-0256]

§82.161 Technician

Certification.

(a) Effective [eighteen months from publication], persons who maintain, service, or repair appliances, except MVACs, and persons who dispose of appliances, except for small appliances, room air conditioners, and MVACs, must be certified by an approved technician certification program as follows:

(1) persons who maintain, service, or repair small appliances as defined in §82.158 (v) must be properly certified as Type I technicians.

(2) persons who maintain, service, or repair high or very high-pressure appliances, except small appliances and MVACs, or dispose of high or very high-pressure appliances, except small appliances and MVACs, must be properly certified as Type II technicians.

(3) persons who maintain, service, or repair low-pressure appliances or dispose of low-pressure appliances must be properly certified as Type III technicians.

(4) persons who maintain, service, or repair low- and high- pressure equipment as described in §82.161 (a) (1), (2) and (3) must be properly certified as Universal technicians.

(5) persons who maintain, service, or repair MVAC-like appliances must either be properly certified as Type II technicians or complete the

training and certification test offered by a training and certification program approved under 40 CFR §82.40.

(b) Test Subject Material.

The Administrator shall maintain a bank of test questions divided into four groups, including a core group and three technical groups. The Administrator shall release this bank of questions only to approved technician certification programs. Tests for each type of certification shall include a minimum of 25 questions drawn from the core group and a minimum of 25 questions drawn from each relevant technical group. These questions shall address the subject areas listed in Appendix D.

(c) Program Approval. Persons may seek approval of any technician certification program, in accordance with the provisions of this paragraph, by submitting to the Administrator at the address in §82.160 (a) verification that the program meets all of the standards listed in Appendix D and the following standards:

(1) Alternative Examinations. Programs are encouraged to make provisions for non-English speaking technicians by providing tests in other languages or allowing the use of a translator when taking the test. If a translator is used, the certificate received

must indicate that translator assistance was required. A test may be administered orally to any person who makes this request, in writing, to the program at least 30 days before the scheduled date for the examination. The letter must explain why the request is being made.

(2) Recertification. The Administrator reserves the right to specify the need for technician recertification at some future date, if necessary, by placing a notice in the Federal Register.

(3) Proof of Certification. Programs must issue individuals a wallet-sized card to be used as proof of certification, upon successful completion of the test. Programs must issue an identification card to technicians that receive a score of 70 percent or higher on the closed-book certification exam, within 30 days. Programs providing Type I certification using the mail-in format, must issue a permanent identification card to technicians that receive a score of 84 percent or higher on the certification exam, no later than 30 days after the program has received the exam and any additional required material. Each card must include, at minimum, the name of the certifying program, and the date the organization became a certifying program, the name of the person certified, the type of certification, a unique

number for the certified person, and the following text:
[Name of person] has been certified as a [Type I, Type II, Type III, and/or Universal, as appropriate] technician as required by 40 CFR Part 82 Subpart F.

(4) The Administrator reserves the right to consider other factors deemed relevant to ensure the effectiveness of certification programs.

(d) If approval is denied under this section, the Administrator shall give written notice to the program setting forth the basis for her or his determination.

(e) If at any time an approved program violates any of the above requirements, the Administrator reserves the right to revoke approval. In such cases, the Administrator or her or his designated representative shall give notice to the organization setting forth the basis for her or his determination.

(f) Authorized representatives of the Administrator may require technicians to demonstrate on the business entity's premises their ability to perform proper procedures for recovering and/or recycling refrigerant. Failure to demonstrate or failure to properly use the equipment may result in revocation of the certificate.

Failure to abide by any of the provisions of this regulation may also result in revocation or suspension of the certificate. If a technician's certificate is revoked, the technician would need to recertify before maintaining, servicing, repairing or disposing of any appliances.

(g) Persons seeking approval of a technician certification program may also seek approval for technician certifications granted previously under the program. Interested persons may submit to the Administrator at the address in §82.160(a) verification that the program met all of the standards of paragraph 82.161(c) and Appendix D, or verification that the program met all of the standards of paragraph 82.161(c) and Appendix D, except for some elements of the test subject material, in which case the person must submit verification that supplementary information on that material will be provided pursuant to Appendix D, section (j).
[Approved by the Office of Management and Budget under the control number 2060-0256]

§82.162 Certification by Owners of Recovery and Recycling Equipment
(a) No later than [90 days after publication of the final rule], or within 20 days of

commencing business for those persons not in business at the time of promulgation, persons maintaining, servicing, or repairing appliances except for MVACs, and persons disposing of appliances except for small appliances and MVACs, must certify to the Administrator that such person has acquired certified recovery or recycling equipment and is complying with the applicable requirements of this rule.
Such equipment may include system-dependent equipment but must include self-contained equipment, if the equipment is to be used in the maintenance, service, or repair of appliances except for small appliances. The owner or lessee of the recovery or recycling equipment may perform this certification for his or her employees. Certification shall take the form of a statement signed by the owner of the equipment or another responsible officer and setting forth:

(1) The name and address of the purchaser of the equipment, including the county name,

(2) The name and address of the establishment where each piece of equipment is or will be located,

(3) The number of service trucks (or other vehicles) used to transport technicians and equipment between the establishment and job sites and the field, and

(4) The manufacturer name, the date of

manufacture, and if applicable, the model and serial number of the equipment.

(5) The certification must also include a statement that the equipment will be properly used in servicing or disposing of appliances and that the information given is true and correct.

Owners or lessees of recycling or recovery equipment having their places of business in:

Connecticut
Maine
Massachusetts
New Hampshire
Rhode Island
Vermont

must send their certifications

to:

CAA §608 Enforcement
Contact
EPA Region I
Mail Code APC
JFK Federal Building
One Congress Street
Boston, MA 02203

Owners or lessees of recycling or recovery equipment having their places of business in:

New York
New Jersey
Puerto Rico
Virgin Islands

must send their certifications to:

CAA §608 Enforcement
Contact
EPA Region II
Jacob K. Javits Federal Building
26 Federal Plaza
Room 5000
New York, NY 10278

Owners or lessees of recycling or recovery equipment having their places of business in:

Delaware
District of Columbia
Maryland
Pennsylvania
Virginia
West Virginia

must send their certifications to:

CAA §608 Enforcement
Contact
EPA Region III
Mail Code 3AT21
841 Chestnut Building
Philadelphia, PA 19107

Owners or lessees of recycling or recovery equipment having their places of business in:

Alabama
Florida
Georgia
Kentucky
Mississippi
North Carolina
South Carolina
Tennessee

must send their certifications to:

CAA §608 Enforcement

Contact
EPA Region IV
345 Courtland Street, NE
Mail Code APT-AE
Atlanta, GA 30365

Owners or lessees of recycling or recovery equipment having their places of business in:

Illinois
Indiana
Michigan
Minnesota
Ohio
Wisconsin

must send their certifications to:

CAA §608 Enforcement
Contact
EPA Region V
Mail Code AT18J
77 W. Jackson Blvd.
Chicago, IL 60604-3507

Owners or lessees of recycling or recovery equipment having their places of business in:

Arkansas
Louisiana
New Mexico
Oklahoma
Texas

must send their certifications to:

CAA §608 Enforcement
Contact
EPA Region VI
Mail Code 6T-EC
First Interstate Tower at
Fountain Place
1445 Ross Ave., Suite

1200
Dallas, TX 75202-2733

Owners or lessees of recycling or recovery equipment having their places of business in:

Iowa
Kansas
Missouri
Nebraska

must send their certifications to:

CAA §608 Enforcement
Contact
EPA Region VII
Mail Code ARTX/ARBR
726 Minnesota Ave.
Kansas City, KS 66101

Owners or lessees of recycling or recovery equipment having their places of business in:

Colorado
Montana
North Dakota
South Dakota
Utah
Wyoming

must send their certifications to:

CAA §608 Enforcement
Contact
EPA Region VIII
Mail Code 8AT-AP
999 18th Street, Suite 500
Denver, CO 80202-2405

Owners or lessees of recycling or recovery equipment having their places of business in:

American Samoa
Arizona
California
Guam
Hawaii
Nevada

must send their certifications to:

CAA §608 Enforcement
Contact
EPA Region IX
Mail Code A-3
75 Hawthorne Street
San Francisco, CA 94105

Owners or lessees of recycling or recovery equipment having their places of business in:

Alaska
Idaho
Oregon
Washington

must send their certifications to:

CAA §608 Enforcement
Contact
EPA Region X
Mail Code AT-082
1200 Sixth Ave
Seattle, WA 98101

(b) Certificates under paragraph (a) of this section are not transferable. In the event of a change of ownership of an entity that maintains, services, or repairs appliances except MVACs, or that disposes of appliances except small appliances, MVACs, and MVAC-like appliances, the new owner of the entity shall certify within 30 days of the change of ownership pursuant to paragraph (a) of this section.

(c) No later than [90 days after publication of the final rule], persons recovering refrigerant from small appliances, MVACs, and MVAC-like appliances for purposes of disposal of these appliances must certify to the Administrator that such person has acquired recovery equipment that meets the standards set forth in §82.158(l) and/or (m), as applicable, and that such person is complying with the applicable requirements of this rule. Such equipment may include system-dependent equipment but must include self-contained equipment, if the equipment is to be used in the disposal of appliances except for small appliances. The owner or lessee of the recovery or recycling equipment may perform this certification for his or her employees. Certification shall take the form of a statement signed by the owner of the equipment or another responsible officer and setting forth:

(1) The name and address of the purchaser of the equipment, including the county name;

(2) The name and address of the establishment where each piece of equipment is or will be located;

(3) The number of service

trucks (or other vehicles) used to transport technicians and equipment between the establishment and job sites and the field; and

(4) The manufacturer name, the date of manufacture, and if applicable, the model and serial number of the equipment.

(5) The certification must also include a statement that the equipment will be properly used in recovering refrigerant from appliances and that the information given is true and correct. The certification shall be sent to the appropriate address in paragraph (a).

(d) Failure to abide by any of the provisions of this subpart may result in revocation or suspension of certification under paragraphs (a) or (c) of this section. In such cases, the Administrator or her or his designated representative shall give notice to the organization setting forth the basis for her or his determination.

[Approved by the Office of Management and Budget under the control number 2060-0256]

§ 82.164 Reclaimer Certification

Effective [90 days after publication of the final rule], persons reclaiming used refrigerant for sale to a new owner must certify to the Administrator that such person will:

(a) return refrigerant to at least the standard of purity set forth in ARI Standard 700-1988, Specifications for Fluorocarbon Refrigerants,

(b) verify this purity using the methods set forth in ARI Standard 700-1988,

(c) release no more than 1.5 percent of the refrigerant during the reclamation process, and

(d) dispose of wastes from the reclamation process in accordance with all applicable laws and regulations.

The data elements for certification are as follows:

(1) The name and address of the reclaimer,

(2) A list of equipment used to reprocess and to analyze the refrigerant, and

(3) The owner or a responsible officer of the reclaimer must sign the certification stating that the refrigerant will be returned to at least the standard of purity set forth in ARI Standard 700-1988, Specifications for Fluorocarbon Refrigerants, that the purity of the refrigerant will be verified using the methods set forth in ARI Standard 700-1988, that no more than 1.5 percent of the refrigerant will be released during the reclamation process, that wastes from the reclamation process will be properly disposed of, and that the information given is true

and correct. The certification should be sent to the following address:

§608 Recycling Program Manager
Stratospheric Protection Division
(6205-J)
U.S. Environmental Protection Agency
401 M Street, S.W.
Washington, D.C. 20460

(e) Certificates are not transferable. In the event of a change in ownership of an entity which reclaims refrigerant, the new owner of the entity shall certify within 30 days of the change of ownership pursuant to this section.

(f) Failure to abide by any of the provisions of this subpart may result in revocation or suspension of the certification of the reclaimer. In such cases, the Administrator or her or his designated representative shall give notice to the organization setting forth the basis for her or his determination.
[Approved by the Office of Management and Budget under the control number 2060-0256]

§82.166 Reporting and Recordkeeping Requirements.

(a) All persons who sell or distribute any class I or class II substance for use as a refrigerant must retain invoices that indicate the name of the purchaser, the date of sale, and the quantity of refrigerant purchased.

(b) Purchasers of any class I or class II refrigerants who employ technicians who recover refrigerants may provide evidence of each technician's certification to the wholesaler who sells them refrigerant; the wholesaler will then keep this information on file. In such cases, the purchaser must notify the wholesaler regarding any change in a technician's certification or employment status.

(c) Approved equipment testing organizations must maintain records of equipment testing and performance and a list of equipment that meets EPA requirements. A list of all certified equipment shall be submitted to EPA within 30 days of the organization's approval by EPA and annually at the end of each calendar year thereafter.

(d) Approved equipment testing organizations shall submit to EPA within 30 days of the certification of a new model line of recycling or recovery equipment the name of the manufacturer and the name and/or serial number of the model line.

(e) Approved equipment testing organizations shall notify EPA if retests of equipment or inspections of manufacturing facilities conducted pursuant to §82.158(j) show that a previously certified model line fails to meet EPA requirements. Such notification must be received within thirty days of the retest or inspection.

(f) Programs certifying technicians must maintain records in accordance with section (g) of Appendix D of this rulemaking.

(g) Reclaimers must maintain records of the names and addresses of persons sending them material for reclamation and the quantity of the material (the combined mass of refrigerant and contaminants) sent to them for reclamation. Such records shall be maintained on a transactional basis.

(h) Reclaimers must maintain records of the quantity of material sent to them for reclamation, the mass of refrigerant reclaimed, and the mass of waste products. Reclaimers must report this information to the Administrator annually within 30 days of the end of the calendar year.

(i) Persons disposing of small appliances, MVACs, and MVAC-like appliances must maintain copies of signed statements obtained pursuant to §82.156(f)(2).

(j) Persons servicing appliances normally containing 50 or more pounds of refrigerant must provide the owner/operator of such appliances with an invoice or other documentation, which indicates the amount of refrigerant added to the appliance.

(k) Owners/operators of appliances normally containing 50 or more pounds of refrigerant must keep servicing records documenting the date and type of service, as well as the quantity of refrigerant added. The owner/operator must keep records of refrigerant purchased and added to such appliances in cases where owners add their own refrigerant. Such records should indicate the date(s) when refrigerant is added.

(l) Technicians certified under §82.161 must keep a copy of their certificate at their place of business.

(m) All records required to be maintained pursuant to this section must be kept for a minimum of three years unless otherwise indicated. Entities that dispose of appliances must keep these records on-site.
    [Approved by the Office of Management and Budget under the control number 2060-0256]

Appendix A--ARI Standard 700-1988

Appendix B--ARI Standard 740-1993

Appendix C--Method for Testing Recovery Devices for Use with Small Appliances

APPENDIX V

FEDERAL REGISTERs CFR 40 Part 82 (Section 609)
And 40 CFR 82 Appendix D (Training)

The following Code of Federal Regulations (CFR) is printed in a larger readable type than the published versions that can be obtained from the Government Printing Office. This Appendix excludes the 40 CFR Part 82 summary and Appendices. The majority of the summary and Appendices material is covered throughout this text. Official copies of the summary and complete regulations can be obtained from the Government Printing Office (GPO) at 202-512-1530 or by calling the EPA Hotline at 1-800-296-1996. If you have a personal computer with a modem and communications program you can download an unofficial copy of 40 CFR Part 82 (file MVAC.ZIP) from the EPA's TTN Clean Air Act electronic bulletin board. The access number is 919-541-5742.

This chapter was downloaded electronically from EPA's bulletin board. The EPA cautions that, "the electronic dissemination of EPA files is an experimental project of EPA and the GPO. While every effort has been made to ensure that these electronic files accurately reflect the contents of the CFR, this **IS NOT AN OFFICIAL VERSION** and it **SHOULD NOT BE RELIED UPON** for purposes of legal enforcement or citation. In the event that the material in the electronic files varies from the text of the CFR, the official version as published by the Administrative Committee of the Federal Register (ACFR) controls." Obtain official copies for your shop from the resources listed above. This CFR copy was published prior to several of the EPA's final decisions.

The author and publisher make no warranties, either expressed or implied, with respect to the information contained herein. This CFR was provided by the EPA bulletin board as mentioned above. The author and publisher shall not be liable for any incidental or consequential damages in connection with, or arising out of, the use of material in this book.

CFR 40 Part 82 (Section 609)

For the reasons set out in the Preamble, EPA is hereby amending 40 CFR Part 82 as follows:

PART 82 - PROTECTION OF STRATOSPHERIC OZONE

1. Authority: The authority citation for part 82 Subpart B is as follows:

Authority: 42 U.S.C. § 7414, § 7601, § 7671 - 7671q.

2. Part 82 is amended by adding Subpart B to read as follows:

Subpart B - Servicing of Motor Vehicle Air Conditioners

82.30 Purpose and Scope
82.32 Definitions
82.34 Prohibitions
82.36 Approved Refrigerant Recycling Equipment
82.38 Approved Independent Standards Testing Organizations
82.40 Technician Training and Certification
82.42 Certification and Recordkeeping Requirements
Appendix A to Part 82 Subpart B - Standard for Recycle/Recover Equipment
Appendix B to Part 82 Subpart B - Standard for Recover Equipment

§ 82.30 Purpose and Scope

(a) The purpose of these regulations is to implement section 609 of the Clean Air Act, as amended (Act) regarding the servicing of motor vehicle air conditioners.

(b) These regulations apply to any person performing service on a motor vehicle for consideration when this service involves the refrigerant in the motor vehicle air conditioner.

§ 82.32 Definitions

(a) Approved Independent Standards Testing Organization means any organization which has applied for and received approval from the Administrator pursuant to §82.38.

(b) Approved Refrigerant Recycling Equipment means equipment certified by the Administrator or an organization approved under § 82.38 as meeting either one of the standards in § 82.36. Such equipment extracts and recycles refrigerant or extracts refrigerant for recycling on-site or reclamation off-site.

(c) Motor vehicle as used in this Subpart means any vehicle which is self-propelled and designed for transporting persons or property on a street or highway, including but not limited to passenger cars, light duty vehicles, and heavy duty vehicles. This definition does not include a vehicle where final assembly of the vehicle has not been completed by the original equipment manufacturer.

(d) Motor vehicle air conditioners means mechanical vapor compression refrigeration equipment used to cool the driver's or passenger's compartment of any motor vehicle. This definition is not intended to encompass the hermetically sealed refrigeration systems used on motor vehicles for refrigerated cargo and the air conditioning systems on passenger buses using HCFC-22 refrigerant.

(e) Properly using means using equipment in conformity with Recommended Service Procedure for the Containment of R-12 (CFC-12) set forth in Appendix A to this Subpart. In addition, this term includes operating the equipment in accordance with the manufacturer's guide to operation and maintenance and using the equipment only for the controlled substance for which the machine is designed. For equipment that extracts and recycles refrigerant, properly using also means to recycle refrigerant before it is returned to a motor vehicle air conditioner. For equipment that only recovers refrigerant, properly using includes the requirement to recycle the refrigerant on-site or send the refrigerant off-site for reclamation. Refrigerant from reclamation facilities that is used for the purpose of recharging motor vehicle air conditioners must be at or above the standard of purity developed by the Air-conditioning and Refrigeration Institute (ARI 700-88) in effect as of November 15, 1990. Refrigerant may be recycled off-site only if the refrigerant is extracted using recover only equipment, and is subsequently recycled off-site by equipment owned by the person that owns both the recover only equipment and owns or operates the establishment at which the refrigerant was extracted. In any event, approved equipment must be used to extract refrigerant prior to performing any service during which discharge of refrig-erant from the motor vehicle air conditioner can reasonably be expected. Intentionally venting or disposing of refrigerant to the atmosphere is an improper use of equipment.

(f) <u>Refrigerant</u> means any class I or class II substance used in a motor vehicle air condi-tioner. Class I and Class II substances are listed in Part 82, Subpart A, Appendix A. Effective November 15, 1995, refrigerant shall also include any substitute substance.

(g) <u>Service for consideration</u> means being paid to perform service, whether it is in cash, credit, goods, or services. This includes all service except that done for free.

(h) <u>Service involving refrigerant</u> means any service during which discharge or release of refrigerant from the motor vehicle air conditioner to the atmosphere can reasonably be expected to occur.

§ 82.34 Prohibitions

(a) Effective August 12, 1992, no person repairing or servicing motor vehicles for consideration may perform any service on a motor vehicle air conditioner involving the refrigerant for such air conditioner

(1) without properly using equipment approved pursuant to § 82.36 and
(2) unless such person has been properly trained and certified by a technician certification program approved by the Administrator pursuant to § 82.40.

The requirements of this paragraph do not apply until January 1, 1993 for small entities who certify to the Administrator in accordance with section §82.42(a)(2).

(b) Effective November 15, 1992, no person may sell or distribute, or offer for sale or distribution, any class I or class II substance that is suitable for use as a refrigerant in a motor vehicle air-conditioner and that is in a container which contains less than 20 pounds of such refrigerant to any person unless that person is properly trained and certified under § 82.40 or intended the containers for resale only, and so certifies to the seller under §82.42(b)(4).

(c) No technician training programs may issue certificates unless the program complies with all of the standards in §82.40(a).

82.36 Approved Refrigerant Recycling Equipment

(a) (1) Refrigerant recycling equipment must be certified by the Administrator or an independent standards testing organization approved by the Administrator under §82.38 to meet either one of the following standards:

(2) Equipment that recovers and recycles refrigerant must meet the standards set forth in Appendix A to this subpart (Recommended Service Procedure for the Containment of R-12, Extraction and Recycle Equipment for Mobile Automotive Air-Conditioning Systems, and Standard of Purity for Use in Mobile Air Conditioning Systems).

(b) Refrigerant recycling equipment purchased before September 4, 1991 that has not been certified under paragraph (a) of this section shall be considered approved if the equipment is substantially identical to equipment certified under paragraph (a) of this section. Equipment manufacturers or owners may request a determination by the Administrator by submitting an application and supporting documents which indicate that the equipment is substantially identical to approved equipment to:

MVACs Recycling Program Manager Stratospheric Ozone Protection Branch
(6202-J)
U.S.Environmental Protection Agency
401 M Street, S.W.
Washington, D.C. 20460
Attn. Substantially Identical Equipment Review

Supporting documents must include process flow sheets, lists of components and any other information which would indicate that the equipment is capable of cleaning the refrigerant to the standards in Appendix A. Authorized representatives of the Administrator may inspect equipment for which approval is being sought and request samples of refrigerant that has been extracted and/or recycled using the equipment. Equipment which fails to meet appropriate standards will not be considered approved.

(c) The Administrator will maintain a list of approved equipment by manufacturer and model. Persons interested in obtaining a copy of the list should send written inquiries to the address in (b) of this section.

§ 82.38 Approved Independent Standards Testing Organizations

(a) Any independent standards testing organization may apply for approval by the Administrator to certify equipment as meeting the standards in Appendix A. The application shall be sent to:

MVACs Recycling Program Manager
Stratospheric Ozone Protection Branch
(6202-J)
U.S. Environmental Protection Agency
401 M Street, S.W.
Washington, D.C. 20460

(b) Applications for approval must document the following:

(1) That the organization has the capacity to accurately test whether refrigerant recycling equipment complies with the applicable standards. In particular, applications must document:

(i) The equipment present at the organization that will be used for equipment testing;

(ii) The expertise in equipment testing and the technical experience of the organization's personnel;

(iii) Thorough knowledge of the standards as they appear in Appendix A of this subpart; and

(iv) The test procedures to be used to test equipment for compliance with applicable standards, and why such test procedures are appropriate for that purpose.

(2) That the organization has no conflict of interest and will receive no financial benefit based on the outcome of certification testing; and

(3) That the organization agrees to allow the Administrator access to verify the information contained in the application.

(c) If approval is denied under this section, the Administrator shall give written notice to the organization setting forth the basis for his or her determination.

(d) If at any time an approved independent standards testing organization is found to be conducting certification tests for the purposes of this subpart in a manner not consist with the representations made in its

application for approval under this section, the Administrator reserves the right to revoke approval.

§ 82.40 Technician Training and Certification

(a) Any technician training and certification program may apply for approval, in accordance with the provisions of this paragraph, by submitting to the Administrator at the address in 82.38 (a) verification that the program meets all of the following standards:

(1) Training.
Each program must provide adequate training, through one or more of the following means: on-the-job training, training through self-study of instructional material, or on-site training involving instructors, videos or a hands-on demonstration.

(2) Test Subject Material.
The certification tests must adequately and sufficiently cover the following:

(i) The standards established for the service and repair of motor vehicle air conditioners as set forth in Appendix A to this subpart. These standards relate to the recommended service procedures for the containment of refrigerant, extraction and recycle equipment, and the standard of purity for refrigerant in motor vehicle air conditioners.

(ii) Anticipated future technological developments, such as the introduction of HFC-134a in new motor vehicle air conditioners.

(iii) The environmental consequences of refrigerant release and the adverse effects of stratospheric ozone layer depletion.

(iv) As of [August 13, 1992, the requirements imposed by the Administrator under §609 of the Act.

(3) Test Administration.

Completed tests must be graded by an entity or individual who receives no benefit based on the outcome of testing; a fee may be charged for grading. Sufficient measures must be taken at the test site to ensure that tests are completed honestly by each technician. Each test must provide a means of verifying the identification of the individual taking the test. Programs are encouraged to make provisions for non-English speaking technicians by providing tests in other languages or allowing the use of a translator when taking the test. If a translator is used, the certificate received must indicate that translator assistance was required.

(4) Proof of Certification.

Each certification program must offer individual proof of certification, such as a certificate, wallet-sized card, or display card, upon successful completion of the test. Each certification program must provide a unique number for each certified technician.

(b) In deciding whether to approve an application, the Administrator will consider the extent to which the applicant has documented that its program meets the standards set forth in this section. The Administrator reserves the right to consider

other factors deemed relevant to ensure the effectiveness of certification programs. The Administrator may approve a program which meets all of the standards in paragraph (a) except test administration if the program, when viewed as a whole, is at least as effective as a program that does meet all the standards. Such approval shall be limited to training and certification conducted before August 13, 1992. If approval is denied under this section, the Administrator shall give written notice to the program setting forth the basis for his determination.

(c) Technical Revisions.

Directors of approved certification programs must conduct periodic reviews of test subject material and update the material based upon the latest technological developments in motor vehicle air conditioner service and repair. A written summary of the review and any changes made must be submitted to the Administrator every two years.

(d) Recertification.

The Administrator reserves the right to specify the need for technician recertification at some future date, if necessary.

(e) If at any time an approved program is conducted in a manner not consistent with the representations made in the application for approval of the program under this section, the Administrator reserves the right to revoke approval.

(f) Authorized representatives of the Administrator may require technicians to demonstrate on

the business entity's premises their ability to perform proper procedures for recovering and/or recycling refrigerant. Failure to demonstrate or failure to properly use the equipment may result in revocation of the technician's certificate by the Administrator. Technicians whose certification is revoked must be recertified before servicing or repairing any motor vehicle air conditioners.

§ 82.42 Certification, Recordkeeping and Public Notification Requirements

(a) Certification requirements

(1) No later than January 1, 1993, any person repairing or servicing motor vehicle air conditioners for consideration shall certify to the Administrator that such person has acquired, and is properly using, approved equipment and that each individual authorized to use the equipment is properly trained and certified. Certification shall take the form of a statement signed by the owner of the equipment or another responsible officer and setting forth:

(i) The name of the purchaser of the equipment,
(ii) The address of the establishment where the equipment will be located; and
(iii) The manufacturer name and equipment model number, the date of manufacture, and the serial number of the equipment. The certification must also

include a statement that the equipment will be properly used in servicing motor vehicle air conditioners, that each individual authorized by the purchaser to perform service is properly trained and certified in accordance with section 82.40, and that the information given is true and correct. The certification should be sent to:

MVACs Recycling
Program Manager
Stratospheric Ozone
Protection Branch
(6202-J)
U.S. Environmental
Protection Agency
401 M Street, S.W.
Washington, D.C.
20460

(2) The prohibitions in § 82.34(a) shall be effective as of January 1, 1993 for persons repairing or servicing motor vehicle air conditioners for consideration at an entity which performed service on fewer that 100 motor vehicle air conditioners in calendar year 1990, but only if such person so certifies to the Administrator no later than August 13, 1992. Persons must retain adequate records to demonstrate that the number of vehicles serviced was fewer than 100.

(3) Certificates of compliance are not transferable. In the event of a change of ownership of an entity which services motor vehicle air conditioners for consideration, the new owner of the entity shall certify within thirty days of the change of ownership pursuant to section 82.42(a)(1).

(b) Recordkeeping requirements

(1) Any person who owns approved refrigerant recycling equipment certified under § 82.36(a)(2) must maintain records of the name and address of any facility to which refrigerant is sent.

(2) Any person who owns approved refrigerant recycling equipment must retain records demonstrating that all persons authorized to operate the equipment are currently certified under section 82.40.

(3) Any person who sells or distributes any class I or class II substance that is suitable for use as a refrigerant in a motor vehicle air conditioner and that is in a container of less than 20 pounds of such refrigerant must verify that the purchaser is properly trained and certified under §82.40. The seller must have a reasonable basis for believing that the information presented by the purchaser is accurate. The only exception to these requirements is if the purchaser is purchasing the small containers for resale only. In this case, the seller must obtain a written statement from the purchaser that the containers are for resale only and indicate the purchasers name and business address. Records required under this paragraph must be retained for a period of three years.

(4) All records required to be maintained pursuant to this section must be kept for a minimum of three years unless otherwise indicated. Entities which service motor vehicle air conditioners for consideration must keep these records on-site.

(5) All entities which service motor vehicle air conditioners for consideration must allow an authorized representative of the Administrator entry onto their premises (upon presentation of his or her credentials) and give the authorized representative access to all records required to be maintained pursuant to this section.

(c) Public Notification
Any person who conducts any retail sales of a class I or class II substance that is suitable for use as a refrigerant in a motor vehicle air conditioner, and that is in a container of less than 20 pounds of refrigerant, must prominently display a sign where sales of such containers occur which states: "It is a violation of federal law to sell containers of Class I and Class II refrigerant of less than 20 pounds of such refrigerant to anyone who is not properly trained and certified to operate approved refrigerant recycling equipment."

Appendix A

[Available on EPA bulletin board]

40 CFR 82 APPENDIX D
TRAINING

This Appendix explaines the testing process for Section 608 in detail including the general areas that are coverd in each exam, the number of questions, passing grades, and how exams are given. The EPA is maintaining tight security over the exam process including on-site inspections of testing sites.

The following Code of Federal Regulations (CFR) Appendix is printed in a larger readable type than the published versions that can be obtained from the Government Printing Office. This Appendix excludes the 40 CFR Part 82 summary and other Appendices. Official copies of the summary and complete regulations can be obtained from the Government Printing Office (GPO) at 202-512-1530 or by calling the EPA Hotline at 1-800-296-1996. If you have a personal computer with a modem and communications program you can download an unofficial copy of 40 CFR Part 82 from the EPA's TTN Clean Air Act electronic bulletin board. The access number is 919-541-5742.

This Appendix was downloaded electronically from EPA's bulletin board. The EPA cautions that, "the electronic dissemination of EPA files is an experimental project of EPA and the GPO. While every effort has been made to ensure that these electronic files accurately reflect the contents of the CFR, this **IS NOT AN OFFICIAL VERSION** and it **SHOULD NOT BE RELIED UPON** for purposes of legal enforcement or citation. In the event that the material in the electronic files varies from the text of the CFR, the official version as published by the Administrative Committee of the Federal Register (ACFR) controls." Obtain official copies for your shop from the resources listed above.

The author and publisher make no warranties, either expressed or implied, with respect to the information contained herein. This CFR was provided by the EPA bulletin board as mentioned above. The author and publisher shall not be liable for any incidental or consequential damages in connection with, or arising out of, the use of material in this book.

Standards for Certifying Programs (Section 608)

a. Test Preparation.

Certification for Type II, Type III and Universal technicians will be dependent upon passage of a closed-book, proctored test, administered in a secure environment, by an EPA-approved certifying program.

Certification for Type I technicians will be dependent upon passage of an EPA-approved test, provided by an EPA-approved certifying program. Organizations providing Type I certification only, may chose either an on-site format, or a mail-in format, similar to what is permitted under the MVACs program.

Each certifying program must assemble tests by choosing a prescribed subset from the EPA test bank. EPA expects to have a test bank with minimum of 500 questions, which will enable the certifying program to generate multiple tests in order to discourage cheating. Each test must include 25 questions drawn from Group 1 and 25 questions drawn from each relevant technical Group. Tests for Universal technicians will include 100 questions (25 from Group 1 and 25 from each relevant technical Group). Each 50-question test represents 10 percent of the total test bank. Questions should be divided in order to sufficiently cover each topic within the Group.

Each certifying program must show a method of randomly choosing which questions will be on the tests. Multiple versions of the test must be used during each testing event. Test answer sheets or (for those testing via the computer medium) computer files must include the name and address of the applicant, the name and address of the certifying program, and the date and location at which the test was administered.

Training material accompanying mail-in Type I tests must not include sample test questions mimicking the language of the certification test. All mail-in material will be subject to review by EPA.

Certifying programs may charge individuals reasonable fees for the administration of the tests. EPA will publish a list of all approved certifying programs periodically, including the fees charged by the programs. This information will be available from the Stratospheric Ozone Protection Hotline.

b. Proctoring.

A certifying program for Type II, Type III and Universal technicians must designate or arrange for the designation of at least one proctor registered for each testing event. If more than 50 people are taking tests at the same time at a given site, the certifying organization must adhere to normal testing procedures, by designating at least one additional proctor or monitor for every 50 people taking tests at that site.

The certification test for Type II, Type III and Universal technicians is a closed-book exam. The proctors must ensure that the applicants for certification do not use any notes or training materials during testing. Desks or work space must be placed in a way that discourages cheating. The space and physical facilities are to be conducive to continuous surveillance by the proctors and monitors during testing.

The proctor may not receive any benefit from the outcome of the testing other than a fee for proctoring. Proctors cannot know in advance which questions are on the tests they are proctoring.

Proctors are required to verify the identity of individuals taking the test by examining photo identification. Acceptable forms of identification include but are not limited to drivers' licenses, government identification cards, passports, and military identification.

Certifying programs for Type I technicians using the mail-in format, must take sufficient measures at the test site to ensure that tests are completed honestly by each technician. Each test for Type I certification must provide a means of verifying the identification of the individual taking the test. Acceptable forms of identification include but are not limited to drivers' licenses numbers, social security numbers, and passport numbers.

c. Test Security.

A certifying program must demonstrate the ability to ensure the confidentiality and security of the test questions and answer keys through strict accountability procedures. An organization interested in developing a technician certification program will be required to describe these test security procedures to EPA.

After the completion of a test, proctors must collect all test forms, answer sheets, scratch paper and notes. These items are to be placed in a sealed envelope.

d. Test content.

All technician certification tests will include 25 questions from Group I. Group I will ask questions in the following areas:

I.   Environmental impact of   CFCs and HCFCs

II. Laws and regulations

III. Changing industry

outlook

Type I, Type II and Type III certification tests will include 25 questions from Group II. Group II will ask questions covering sector-specific issues in the following areas:

IV. Leak detection

V. Recovery Techniques

VI. Safety

VII. Shipping

VII. Disposal

Universal Certification will include 75 questions from Group II, with 25 from each of the three sector-specific areas.

e. Grading.

Tests must be graded objectively. Certifying programs must inform the applicant of their test results no later than 30 days from the date of the test. Type I certifying programs using the mail-in format, must notify the applicants of their test results no later than 30 days from the date the certifying programs received the completed test and any required documentation. Certifying programs may mail or hand deliver the results.

The passing score for the closed-book Type I, Type II, Type III and Universal certification test is 70 percent. For Type I certification tests using the mail-in format, passing score is 84 percent.

f. Proof of Certification.

Certifying programs must issue a standard wallet-sized identification card no later than 30 days from the date of the test. Type I certifying programs using mail-in formats must issue cards to certified technicians no later than 30 days from the date the certifying program receives the completed test and any required documentation.

Each wallet-sized identification card must include, at a minimum, the name of the certifying program including the date the certifying program received EPA approval, the name of the person certified, the type of certification, a unique number for the certified person and the following text:

> [name of person] has been certified as [Type I, Type II, Type III and/or Universal -- as appropriate] technician as required by 40 CFR 82 subpart F.

g. Recordkeeping and Reporting Requirements.

Certifying programs must maintain records for at least three years which include but are not limited to the names and addresses of all individuals taking the tests, the scores of all certification tests administered, and the dates and locations of all tests administered.

Certifying programs must send EPA an activity report every six months, the first to be submitted six months following approval of the program by EPA. This report

will include the pass/fail rate and testing schedules. This will allow the Agency to determine the relative progress and success of these programs. If the certifying program believes a test bank question needs to be modified, information about that question should also be included.

Approved certifying programs will receive a letter of approval from EPA. Each testing center must display a copy of that letter.

h. Additional Requirements.

EPA will periodically inspect testing sites to ensure compliance with EPA regulations. If testing center discrepancies are found, they must be corrected within a specified time period. If discrepancies are not corrected, EPA may suspend or revoke the certifying programs's approval. The inspections will include but are not limited to a review of the certifying programs' provisions for test security, the availability of space and facilities to conduct the administrative requirements and ensure the security of the tests, the availability of adequate testing facilities and spacing of the applicants during testing, a review of the proper procedures regarding accountability, and that there is no evidence of misconduct on the part of the certifying programs, their representatives and proctors, or the applicants for certification.

If the certifying programs offer training or provide review materials to the applicants, these endeavors are to be considered completely separate from the administration of the certification test.

i. Approval Process.

EPA anticipates receiving a large number of applications from organizations seeking to become certifying programs. In order to certify as many technicians as possible in a reasonable amount of time, EPA will give priority to national programs. Below are the guidelines EPA will use:

First:

Certifying programs providing at least 25 testing centers with a minimum of one site in at least 8 different states will be considered.

Second:

Certifying programs forming regional networks with a minimum of 10 testing centers will

be considered.

Third:

Certifying programs providing testing centers in geographically isolated areas not sufficiently covered by the national or regional programs will be considered.

Fourth:

All other programs applying for EPA approval will be considered.

j. Grandfathering.

EPA will grandfather technicians whose programs seek and receive EPA approval as a certifying program. As part of this process, these certifying programs may be required to send EPA-approved supplemental information or provide additional testing to ensure the level of the technicians' knowledge. The

certifying programs will also issue new identification cards meeting the requirements specified above.

Persons who are currently technicians must be certified by (18 months from the date of publication). Technicians that participated in certification programs which do not become EPA-approved certifying programs must either receive EPA-approved supplemental information from the original testing organization or be certified by taking a test given by an EPA-approved certification organization by (18 months from the date of publication).

k. Sample Application

EPA has provided a sample application. The Agency designed the application to demonstrate the information certifying programs must provide to EPA. Programs are not required to use this form or this format.

[Approved by the Office of Management and Budget under the control number 2060-0256]

APPENDIX VI

ADDITIONAL RESOURCES

The following EPA summaries, directories, fact sheets and miscellaneous resources are compiled from various EPA publications. Official copies of this material and future updates can be obtained from the Government Printing Office (GPO) at 202-512-1530 or by calling the EPA Hotline at 1-800-296-1996. If you have a personal computer with a modem and a communications program you can download various updates from the EPA's TTN Clean Air Act electronic bulletin board. The access number is 919-541-5742. You can also access the **FEDWORLD National Technical Information Service (NTIS)** electronic bulletin board. From FEDWORLD you can access more than 100 federal bulletin board systems operated by the U.S. Government, including EPA's Office of Research & Development. To connect to FEDWORLD, use your computer communication software to dial FEDWORLD at 703/321-8020. Set your parity to NONE, Data Bits to 8 and Stop Bit to 1 (N,8,1). Set your terminal emulation to ANSI or VT-100. After making a connection access bulletin board # 94 (ORDBBS (EPA).

Parts of this chapter were downloaded electronically from EPA's bulletin boards. The EPA cautions that, "the electronic dissemination of EPA files is an experimental project of EPA and the GPO. While every effort has been made to ensure that these electronic files accurately reflect the contents of the CFR, this **IS NOT AN OFFICIAL VERSION** and it **SHOULD NOT BE RELIED UPON** for purposes of legal enforcement or citation. In the event that the material in the electronic files varies from the text of the CFR, the official version as published by the Administrative Committee of the Federal Register (ACFR) controls." Obtain official copies for your shop from the resources listed above.

The sample questions in this book are referenced to these resources, regulations, reference appendices, and key chapters. To best comprehend the regulations and prepare for the certification test, read the sample questions and then go to the specific reference and read the reference completely. This will help you retain the information.

The author and publisher make no warranties, either expressed or implied, with respect to the information contained herein. This data was provided by the EPA bulletin board or through official EPA memorandums, bulletins, and notices. The author and publisher shall not be liable for any incidental or consequential damages in connection with, or arising out of, the use of material in this book.

EPA PROGRAM HOTLINES

RCRA/SUPERFUND	800-424-9346
SOLID WASTE	800-424-9346
UNDERGROUND STORAGE TANKS	800-424-9346
EMERGENCY PLANNING AND COMMUNITY RIGHT-TO-KNOW	800-535-0202
NATIONAL SMALL FLOWS CLEARINGHOUSE/ WASTEWATER PROGRAMS	800-624-8301
OIL POLLUTION ACT/SPCC	202-260-2342
GROUNDWATER PROTECTION	800-426-4791
SAFE DRINKING WATER HOTLINE	800-426-4791
WETLANDS PROTECTION	800-832-7828
TOXIC SUBSTANCES CONTROL ACT (TSCA) ASSISTANCE	202-554-1404
STORM WATER	703-821-4823
INDOOR AIR QUALITY CLEARINGHOUSE	800-438-4318
ACID RAIN HOTLINE	202-233-9620

STRATOSPHERIC OZONE INFORMATION	800-296-1996
GREENLIGHTS HOTLINE	202-775-6650
AIR RISC HOTLINE (RISK INFORMATION SUPPORT CENTER)	919-541-0888
POLLUTION PREVENTION CLEARINGHOUSE	202-260-1023
NATIONAL RADON HOTLINE	800-767-7236
PUBLIC INFORMATION CENTER	202-260-7751
CONTROL TECHNOLOGY CENTER	919-541-0800

SIGNIFICANT NEW ALTERNATIVES POLICY
SNAP

Section 612 of the Clean Air Act Amendments of 1990 requires EPA to publish lists of acceptable and unacceptable substitutes for ozone depletion substances (ODSs). To complete these lists, the Significant New Alternatives Policy (SNAP) program is evaluating both new and existing alternatives to ODSs on the basis of ODP, GWP, flammability, toxicity, exposure, and economic and technical feasibility within a comparative risk framework. Substitutes that pose greater overall risk to human health and the environment relative to other available alternatives will be restricted. Substitutes are defined as chemical products, or alternate manufacturing processes and are evaluated by the use application. Through these evaluations, SNAP will generate lists of acceptable and unacceptable substitutes for each of the major use sectors: refrigeration/air conditioning, solvent cleaning, fire extinguishing, foam blowing, aerosols, sterilization, tobacco expansion, pesticides, and adhesives, coatings, and inks. (Other uses may fall under the small use exception). Initial decisions on key substitutes have been made, through the NPRM is still under review. Questions about specific alternatives can be addressed to the SNAP Coordinator.

What People will have to do to be in compliance.

- Manufacturers, Importers, Formulators, and Processors must submit information about substitutes for review through SNAP's 90-day notification program.

- It will be illegal to replace an ODS with a substitute listed by SNAP as "unacceptable".

EPA's Refrigerant contact is Rosemary Workman, (202) 233-9727.

EPA Stratospheric Ozone Protection

Resources   Air and Radiation (6205J)   EPA-430-F-93-005   Revised 9/93

AIR CONDITIONING AND REFRIGERATION

This resource directory has been produced to assist people who are responsible for making decisions about air conditioning and refrigeration systems and how to address the January 1, 1996 phaseout of the CFC refrigerants. Additional information can be obtained by contacting the EPA Hotline and the companies and trade associations listed here. Acting now is essential to avoid the costs and possible delays of retrofitting or replacing refrigeration and air conditioning equipment that relies on CFCs. If owners wait to act, they will discover that equipment demand exceeds supply, costs skyrocket, and enforcement is serious and expensive.

The Environmental Protection Agency operates a hotline to provide information to the public on ozone depletion issues. Updated information on the regulations being developed under Title VI of the Clean Air Act Amendments, including information on refrigerant substitutes, is available by calling this toll-free number.

Stratospheric Ozone Hotline   (800) 296-1996

Trade and Professional Associations

Air Conditioning Contractors of America
Ellen Larson   (202) 483-9370

Air Conditioning and Refrigeration Institute
Ed Dooley   (703) 524-8800

Alliance for Responsible CFC Policy
Dave Stirpe   (703) 243-0344

American Gas Cooling Center
Rich Sweetser   (703) 841-8411

American Society of Heating, Refrigerating and
Air Conditioning Engineers
Technical Services   (404) 636-8400

American Subcontractors Association
Sarah Thomson   (703) 684-3450

American Supply Association
Public Affairs Office   (312) 464-0090

Association of Energy Engineers
Ruth Bennett   (404) 447-5083

Building Owners and Managers Association
Michael Jawer   (202) 408-2662

Chemical Manufacturers Association
CHEMTREC   (800) 262-8200

(Referrals to chemical manufacture for Technical Info.)

Commercial Refrigerator Manufacturers Association
Tim Rugh   (202) 857-1145

Electric Power Research Institute
Information Hotline   (415) 855-2411
Commercial Bldg. A/C Ctr.   (608) 262-8223
(Most EPRI services should be solicited through your local electric utility)

Food Marketing Institute
Public Affairs Office   (202) 452-8444

International Association of Cold Storage Contractors
Public Affairs Office   (202) 452-1781

International Facility Management Association
Information Services   (713) 6234362

International Institute of Ammonia Refrigeration
Kent Anderson   (202) 857-1110

Mechanical Service Contractors of America
Barbara Dolim   (301) 869-5800

National Association of Convenience Stores
Public Affairs Office   (703) 836-4564

National Association of Plumbing, Heating and Cooling Contractors
Joanne Oxley   (800) 533-7694

Refrigerating Engineers and Technicians Assn.
Public Affairs Office   (312) 644-6610

Refrigeration Service Engineers Society
Dean Lewis   (708) 297-6464

Sheet Metal and Air Conditioning Contractors
National Association
Public Affairs Office   (703) 803-2980

Air Conditioning Manufacturers

(Note: The following U.S. companies make air conditioning and refrigeration equipment for commercial applications that do not require the use of CFCs.)

Chillers:

Carrier Corporation	(315) 433-4376
Dunham-Bush, Inc.	(203) 249-8671
SnyderGeneral Corporation	(612) 533-5330
The Trane Company	(608) 787-2000
York International Corp.	(717) 771-7890

Rooftop Equipment:

Aaon, Inc.	(918) 583-2266
Carrier Corporation	(315) 432-6000
Goettl Air Conditioning	(602) 275-1515
Goodman Mfg. Corp.	(713) 861-2500
Heat Controller	(517) 787-2100
Hupp Industries, Inc.	(216) 851-6200
Inter-City Products	(615) 793-0450
Lennox International	(214) 497-5000
Mammoth, A Nortek Co.	(612) 559-2711
Miller-Picking Corp.	(814) 479-4023
Rheem Mfg. Corp.	(501) 646-4311
SnyderGeneral Corp.	(214) 754-0500
The Trane Company	(608) 787-2000
York International Corp.	(717) 771-7890

Commercial Refrigeration Manufacturers

Barker Company, Ltd.	(319) 293-3777
Charles L. Frank & Associates	(516) 563-2255
Columbus Show Case Company	(614) 299-3161
Displaymor Manufacturing Co.	(310) 323-5223
Federal Industries	(608) 424-3331
Harford Systems (Duracool Div.)	(410) 272-3400
Hill Refrigeration Corp.	(609) 599-9861
Hussman Corporation	(314) 291-2000
Kysor-Warren	(404) 483-5600
Nax of North America	(515) 244-5326
Pinnacle Equipment Corp.	(215) 944-7611
Regal Custom Fixtures Co.	(609) 261-3323
Royal Store Fixture Company	(215) 467-3700

Southern Equipment Company	(314) 481-0660
Tyler Refrigeration Corp.	(616) 683-2000
Zero Zone Refrigerator Mfg.	(414) 547-0055

(Note.- These are not necessarily complete listings in each *category. EPA does not endorse the companies listed here. For a more complete listing, contact the Air Conditioning and Refrigeration Institute or the Commercial Refrigerator Manufacturers Association.)*

Chemical Manufacturers
(Note: The following companies are known to produce alternative refrigerants.)

Allied-Signal, Inc.	(800) 631-8138
E.I. DuPont de Nemours & Co.	(800) 441-9442
Elf Atochem, N.A.	(800) 343-7940
FMC Lithium Division	(800) 362-2549
Great Lakes Chemical Corp.	(317) 497-6100
ICI Americas, Inc.	(800) 243-5532
Laroche Chemicals	(800) 248-6336

Selected Written Materials

EPA Fact Sheets

The Environmental Protection Agency's Stratospheric Protection Division has produced fact sheets on a number of related issues:

- Section 608: Complying with the Refrigerant Recycling Rule
- Section 606: The Accelerated Phaseout of Ozone - Depleting Substances
- Cooling and Refrigerating Without CFCs
  Short List of Substitute Refrigerants

Copies are available through the Hotline at (800) 296-1996.

Related Federal Register Notices

Production Phaseout and Controls
Final Rule - 57 FR 33754-33798, July 30, 1992

Accelerated Phaseout
Proposed Rule - 58 FR 15014-15049, March 18, 1993

National Recycling and Emissions Reduction Program *Final Rule* - 58 FR 28660-28734, May 14, 1993

General information on the problem of ozone depletion is available by contacting the Stratospheric Ozone Information Hotline. The National Oceanic and Atmospheric Administration issued a report titled 'Our Ozone Shield' on the importance of the ozone layer.

DEFINITIONS

This list of definitions has been extracted from Federal and private sources where appropriate. It is not comprehensive and is only for the purpose of clarifying terminology used in this book.

ARI — The Air-conditioning and Refrigeration Institute with headquarters in Washington, D.C.

ASHRAE — The American Society of Heating, Refrigerating & Air Conditioning Engineers, Inc., with headquarters in Atlanta, Georgia.

Brazed Joint — A gas-tight joint obtained by joining metal parts with alloys that melt at temperatures higher than 800 degrees F (430 degrees C) but less than the melting temperatures of the joined parts.

Carbon Tetrachloride — Used extensively in the United States as a solvent and grain fumigant, and is still used in this capacity in many parts of the world. Carbon tetrachloride is still used as a feedstock in the United States and, therefore, has been identified as a class I substance under Title VI--Stratospheric Ozone Protection--of the Clean Air Act. However, its high toxicity led to a ban of its use in the United States in most dispersive applications.

Chillers — Heavy duty air conditioning systems in commercial and industrial buildings (e.g., air route traffic control centers). There are three types of chillers (reciprocating, screw, and centrifugal) distinguished principally by their compressors. Reciprocating compressors use pistons and cylinders for compression. Screw compressors most commonly use two intermeshing "screws" for compression. As they turn, the volume between the screws is reduced, compressing the refrigerant. Centrifugal compressors rotate at high speed, compressing refrigerant by centrifugal force.

Chlorofluorocarbons (CFC) — Are extremely stable, nontoxic, nonflammable, noncorrosive, and thermally efficient chemicals that are widely used as coolants for refrigeration and air conditioning systems, cleaning agents for electronic components, and foam blowing agents. CFC's are fully halogenated (no hydrogen remaining) halocarbons containing chlorine, fluorine, and carbon atoms.

Class I Substances — Any CFC, halons, carbon tetrachloride, and methylchloroform deemed to fall in this category by the EPA Administrator based on current scientific data and pursuant to the Montreal Protocol, CAA, and EPA's implementing regulations. A complete list of class I substances is contained in Appendix 2-CFC's and CFC Alternatives.

Class 11 Substances — A wide variety of hydrochlorofluorocarbons (HCFC) considered by the EPA Administrator to fall within this category based on current scientific data and in compliance with the Montreal Protocol, CAA, and EPA implementing regulations. A complete list of class II substances is contained in appendix 2.

GWP — Global warming scale.

Halons — Fully halogenated compounds that are effective fire extinguishing chemicals. They are electrically nonconductive, dissipate quickly, leave no residue, are explosive suppressants, and are nontoxic.

Hazardous Waste — Defined in 40 CFR, Part 261.3. A waste is any solid, liquid, or contained gaseous material that is no longer used and is recycled or stored until there is enough time to treat or dispose of it properly. This waste becomes hazardous by virtue of being listed on EPA designated lists and/or having one or more of the following characteristics: ignitability, corrosivity, reactivity, or toxicity.

Hydrochlorofluorocarbons (HCFC) — Types of CFC's that contain hydrogen atoms. Hydrogen reduces the stability of the CFC, allowing the CFC to break down more readily before reaching the stratosphere where it can damage the ozone. HCFC's also contain fluorine, chlorine, and carbon atoms.

Hydrofluorocarbons (HFC) — Halocarbons that contain only fluorine, carbon, and hydrogen.

Methylchloroform (1,1,1-trichloroethane) — Widely used throughout the world as art industrial solvent.

Unlike other class I substances, it is only partially halogenated and correspondingly has a much lower ozone depletion potential (ODP). However, because of its high volume of use, it contributes significantly to total atmospheric chlorine levels.

Motor Vehicle — Any self-propelled vehicle designed for transporting persons or property on a street or highway.

ODP (Ozone Depletion Potential) — The rate of which a chemical destroys the ozone layer. Chemicals are assigned an ODP relative weight to refrigerants and halons. The higher the ODP weight the greater destruction caused by that chemical to the ozone layer. Halon-1211 has an ODP weight of 10 and R-123 (HCFC) has a weight of .016.

Ozone Depletion — The interruption of the naturally occurring ozone generation process. For instance, this occurs when CFCs and halons are released and rise into the stratosphere. Sunlight breaks down the CFC molecule, releasing a chlorine atom, or a bromine atom in the case of halons. Instead of a single oxygen atom combining with the oxygen molecule, the more chemically aggressive chlorine or bromine ions react with an oxygen atom to form chlorine monoxide or another compound which fails to block dangerous ultraviolet radiation. As this process continues, the ozone layer deteriorates, allowing more ultraviolet radiation to pass through and reach the earth's surface.

Ozone Layer — Located 11 miles above the earth's surface and extends beyond 25 miles. Ozone molecules are continually generated as sunlight reacts with oxygen molecules to produce two single oxygen atoms. An oxygen molecule will then combine with a single oxygen atom to produce an ozone molecule. This process is balanced by a simultaneous reaction of ozone decomposing, due to sunlight, into an atom and molecule each of oxygen.

Purging — The removal of noncondensable gases from the cooling system.

Purging Device — An automatic, semi automatic or hand-operated device which collects noncondensable gas from the condenser or receiver, condenses some of the condensable refrigerant, and relieves the remainder to the atmosphere.

Reclaim — To reprocess refrigerant to new conditions by means which may include distillation. It may require chemical analysis of the contaminated refrigerant to determine that appropriate process specifications are met. This term usually implies the use of processes or procedures available only at a reprocessing or manufacturing facility.

Recovery — To remove refrigerant in any condition from a system and store it in an external container without necessarily testing or processing it in any way.

Recovery Equipment — Normally a mechanical system consisting of an evaporator, oil separator, compressor, and condenser which draws refrigerant out of the refrigeration system and stores it in a storage container. The equipment may employ replaceable core filter driers to remove moisture, acid, particulates, and other contaminants.

Recycle — To clean refrigerant for reuse by oil separation and single or multiple passes through moisture absorption devices, such as replaceable core filter-driers, which reduce moisture, acidity, and particulate matter. This term usually implies procedures implemented at the field job site or at a local service shop.

Solid Waste — Defined by the Resource Conservation and Recovery Act (RCRA), section 1004(27) as "discarded material including solid, liquid, semi-solid, or contained gaseous material resulting from industrial, commercial, mining, and agricultural operations, and from community activities." Under this definition, contained gases, such as CFC's and HCFC'S, are clearly solid wastes under RCRA and subject to the regulatory requirements of this Act. On the other hand, uncontained gases, not associated with solid waste management units, are outside of RCRA. However, as of February 5, 1991, EPA suspended these requirements for refrigerants, which exhibit these characteristics and which are recycled, for fear that they might otherwise encourage venting as a means of avoiding this responsibility. EPA is currently studying the issue of CFC's as solid and hazardous waste. See Appendix 1.

DISPOSAL OF CFC's AND CFC-CONTAMINATED MATERIAL

RCRA APPLICABILITY[1]

a. Regulations promulgated pursuant to Subtitle C of the Resource Conservation and Recovery Act (RCRA) (42 U.S.C., 690 1) apply to any discarded materials that are solid wastes (including solids, liquids, semisolids, and contained gases), as defined in 40 CFR 261.2. Contained gases being discarded, including used refrigerants, are considered spent materials and are solid wastes subject to RCRA regulations according to 40 CFR 261.2.

b. Thus, the disposal of CFCs is governed under the framework of the RCRA "cradle to grave" system. RCRA regulations are designed to provide control of hazardous waste by imposing requirements on generators and transporters of hazardous wastes, as well as upon owners and operators of treatment, storage, and disposal facilities (TSDF).

c. RCRA was amended by the Hazardous and Solid Waste Amendments of 1984 (HSWA). These amendments made far-reaching changes to the RCRA regulatory program. Significant new requirements include the Land Disposal Restriction (LDR) regulations and newly identified hazardous wastes. Hazardous waste and material regulations are established in CFR Titles 40 and 49; the occupational safety regulations are set forth in CFR Title 29.

d. Individuals engaged in the disposal of CFC's or CFC-containing substances shall follow RCRA regulations contained in 40 CFR Parts 260-268. See AEE-20 Hazardous Property Management Manual for guidance in managing hazardous materials and hazardous wastes.

e. When disposing of CFCs or CFC-containing substances, individuals shall do so through an EPA-permitted TSDF. If CFCs are classified as hazardous waste, then stricter disposal regulations apply and the CFCs must be assigned an EPA waste code before they can be transported off-site (40 CFR 262.12).

f. However, disposal shall only be considered as a final option when the recapture and recycling/reuse of the CFCs, or the trading and selling of CFCs, is no longer a viable alternative.

g. Given the increasing scarcity of CFCs, one option will be to sell or trade CFCs in the waste exchange market or to sell them to users who have a demand for them. Records of sales of CFCs shall be kept on file indefinitely.

CFCs AS HAZARDOUS WASTE

a. The federal regulations list more than 400 wastes as hazardous (40 CFR 261, Subpart D). These wastes are broken down into four lists, U, P, K, and F (40 CFR 261.33(f), .33(e), .32 and .31). Whenever a waste is generated, the operator must review the lists to determine whether the waste is listed as hazardous.

b. Under the current hazardous waste identification and listing regulations (40 CFR 261), a CFC waste is hazardous only under the circumstances outlined. Four cases exist where a CFC waste would be considered hazardous:

(1) When dichlorofluoromethane (CFC-12) or trichloromonofluoromethane (CFC-11) is an unused commercial chemical product or an off-specification commercial chemical product (including inner liners, containing residues, or spill residues), the material is considered a hazardous waste when discarded, except when sent off-site for recycling. For the purposes of this subsection, the term "unused" means not introduced into a process, activity, or piece of equipment for use. The term "off-specification" shall mean not meeting the physical or chemical standards set by the product manufacturer. The EPA listings for CFC-12 and CFC-11 are U075 and U121, respectively.

---

[1]U.S. Department of Transportation's Chlorofluorocarbons and Halon Use.

(2) When a CFC waste is covered by a spent solvent listing (FOOI-FO05), the waste is considered hazardous. Any CFC solvent used for degreasing would be considered a hazardous waste. Trichloromonofluoromethane (CFC-11) and 1,1,2-trichloro-1,2,2 trifluoromethane used as solvents are considered hazardous wastes. Furthermore, any spent solvent mixture containing CFC's and meeting one of the FOOI-FOO5 solvent listings is considered a hazardous waste. For example, a spent solvent mixture containing 10 percent trichlorofluoromethane, 5 percent ethyl ether, and 5 percent acetone before use would meet the F002 and F003 listings.

(3) If a CFC waste exhibits a characteristic of a hazardous waste (i.e., ignitability, toxicity, reactivity, and corrosivity, see 40 CFR 261.21-261.24), the waste would be considered hazardous. However, if a CFC waste is hazardous by virtue of the toxicity characteristic and is destined for recycling, it is exempt from RCRA regulations (40 CFR 261.4 (b) (I 0- 1 2).

(4) Finally, if a CFC waste is mixed with a hazardous waste, the entire mixture would be a hazardous waste.

c. Filters that are used in the recycling process shall be considered hazardous waste upon removal from the recycling unit. Contamination levels shall be determined as stipulated below.

d. In order to determine the level and type of contamination to recycling filters, an initial laboratory test shall be performed on a sample filter or filters. This need not be conducted on each and every filter used over the life of the equipment. Rather, once an initial baseline test has been performed to determine filter contamination levels, extrapolation from these results may be conducted to estimate future filter contamination levels. This approach can only be followed if system filters are replaced in a consistent and routine fashion. These results can only apply to a specific system in a given geographic area. If filters are not replaced on a regular basis, then the results of the baseline test will not be valid for future filter replacement. A log of filter replacement shall be kept in order to document this approach for regulatory agencies, such as the EPA, should any questions or concerns arise.

e. When a used CFC refrigerant is determined to be a hazardous waste, the owner of the refrigerant system from which it was removed would be considered the generator. In addition, the service person or company that removed the refrigerant from the system would be considered a co-generator. Although parties are subject to RCRA hazardous waste regulations, EPA prefers that the generator responsibility lie with one party, preferably specified in a contract or written agreement.

f. When a CFC refrigerant or solvent is determined to be hazardous waste, it shall be subject to the provisions of the Resource Conservation and Recovery Act (RCRA) and EPA's implementing regulations (40 CFR Parts 124, 261-65, and 270). See the AEE-20 Hazardous Property Management Manual for guidance in meeting these regulatory requirements.

EPA   Stratospheric Ozone Protection

Fact Sheet   Air & Radiation (6205J)   EPA-430-F-93-004 Revised September 1993

SHORT LIST OF ALTERNATIVE REFRIGERANTS

Under the Clean Air Act Amendments of 1990, the U.S. Environmental Protection Agency is required to evaluate substitutes for ozone depleting substances. Both new and existing alternatives will be reviewed and evaluated through EPA's Significant New Alternatives Policy (SNAP) program on the basis of a substance's ozone depletion potential (ODP), global warming potential (GWP), flammability, toxicity, exposure potential, and economic and technical feasibility. Substitutes that pose greater overall risk to human health and the environment relative to other available alternatives will be restricted. Under the SNAP program, EPA will generate lists of acceptable and unacceptable substitutes for each major use sector and a list of substitutes for which requests for approval are pending. EPA published the proposed SNAP rule in the Federal Register on May 12, 1993 (vol. 58, 00, 28094-28192). Promulgation of the final SNAP rule is expected in early 1994. Quarterly updates of the SNAP lists will also be published in the Federal Register.

The following short lists present a select group of proposed acceptable alternatives suitable for chiller and commercial refrigeration retrofits and new equipment. The applicability of a particular alternative may depend on the specifics of the proposed retrofit or replacement. For more current information about the status of the SNAP program and specific refrigerant alternatives, call the Stratospheric Ozone Protection Hotline at 800-296-1966 or the Coordinator at 202-233-9195.

PROPOSED ACCEPTABLE REFRIGERANT ALTERNATIVES FOR CHILLERS

Applications	Alternatives for Retrofitting Equipment			Alternatives for New Equipment				
	HCFC 123	HCFC 22/ HFC 152a/ HCFC 124 blends	HFC 134a	Ammonia systems*	HCFC 123	HCFC 22	HFC 134a	Lithium/ bromide/ water absorption blends
CFC 11 Centrifugal	•			•	•	•	•	•
CFC 12 Centrifugal			•	•	•	•	•	•
CFC 12 Reciprocating			•			•	•	
CFC 500 Centrifugal		•	•	•	•	•	•	•

*ASHRAE Standard 15 - 1992 sets special requirements for systems using ammonia-containing refrigerants.

PROPOSED ACCEPTABLE REFRIGERANT ALTERNATIVES FOR REFRIGERATION EQUIPMENT

Applications	Alternatives for Retrofitting Equipment				Alternatives for New Equipment			
	HCFC 22	HCFC 22/ HFC 152a/ HCFC 124 blends	HCFC 22/ propane/ HFC 125 blends	HFC 134a	Ammonia systems*	HCFC 22	HCFC 22/ propane/ HFC 125 blends	HFC 134a
CFC 12 Cold Storage Warehouses	•	•		•	•	•		•
CFC 12 Commercial Ice Machines		•			•	•		•
CFC 12 Industrial Process Refrigeration	•	•		•	•	•		•
CFC 12 Refrigerated Transport		•		•		•		•
CFC 12 Retail Food	•	•		•	•	•		•
CFC 12 Vending Machines	•	•		•		•		•
CFC 500 Refrigerated Transport		•	•	•		•	•	•
CFC 502 Cold Storage Warehouses	•		•		•	•	•	
CFC 502 Industrial Process Refrigeration	•		•	•	•	•	•	•
CFC 502 Refrigerated Transport	•		•	•		•	•	•
CFC 502 Retail Food	•		•		•	•	•	
CFC 502 Commercial Ice Machines					•	•		

*ASHRAE Standard 15-1992 sets special equipment room requirements for systems using ammonia-containing refrigerants.

EPA Stratospheric Ozone Protection
Final Rule Summary   Air and Radiation (6205J)   EPA-430-F-93-004   9/1993
MEETING LABELING REGULATION REQUIREMENTS

This guidance offers manufacturers, distributors, wholesalers, and retailers instruction on how to comply with the final labeling regulation. Examples illustrate many of the key regulatory features and clarifies the issues that have been raised since the rule's publication on February 11, 1993 (58 FR 8136).

KEY FEATURES OF THE FINAL REQUIREMENTS

On November 15, 1990, Congress amended the Clean Air Act (CAA). Section 611 of the Act as amended, requires labeling of products made with or containing class I and class 11 ozone-depleting substances. It also requires that containers containing class I or class II substances be labeled. The final regulation includes the following key requirements:

1) Treatment of Products and Imports Manufactured Prior to May 15, 1993

All products made before May 15, 1993 are exempt from the labeling requirements if the manufacturer is able to show *within 24 hours,* upon request that its products were made before that date.

If an importer imports products Made prior to May 15, 1993, he/she MUST be able to show, upon request by EPA, that such products were made before the deadline. The date of manufacture may appear on supplemental printed material such as shipping papers, bills of lading, and invoices, or may be made available through index code references, or any other means by which a company tracks its products.

2) Products Manufactured with Class I Substances

Label Pass-Through Requirement
Manufacturers of products that use a class I substance must label their products. Such products manufactured with class I substances may be electronic parts washed in class I solvents, such as electrical components and metal products, plumbing fixtures, and products using class I adhesives, such as some packaging, books, and sporting goods. If a manufacturer purchases a product from a supplier that labels its product "manufactured with," the manufacturer does not need to incorporate that information into a label on its final

product. in other words, manufacturers need only label their products according to their *own direct* manufacturing process. Labels on products containing class I substances and containers of class I or class II substances, however, must be passed through the stream of commerce to the ultimate consumer, since the ozone-depleting substance is contained at the time of purchase. In the case of adhesive or solvent products, the purchaser is likely to release the substance upon application of the product.

For example, a product containing, such as an adhesive, must be labeled as "containing." When that product is applied by a subsequent manufacturer in affixing a cushion to a seat, the seat must be labeled as a "product manufactured with" because the CFCs have been released. The subsequent sale of the seat to an automobile manufacturer would not result in the labeling of a car based on that product

Subsidiaries and the Label Pass-Through Requirement
The rule states that wholly-owned subsidiaries are part of a parent company and are required to pass the warning statement between subsidiaries. If a subsidiary is not 100 percent wholly-owned, the label is not required to be passed through from one subsidiary to another.

If a parent company owns 100 percent of another company and sells a small portion following the effective date of this regulation, EPA may look unfavorably upon the parent company if it appears that the company made the change with the intention of avoiding the label pass-through requirement by selling a small share of its subsidiary company.

*A Reduction in Use of CFC-113 and/or
Methyl Chloroform Over 1990 Use*
If a company (including its divisions, branches, or facilities) has achieved a total use reduction of methyl chloroform (MCF) and/or CFC-113 used as solvents in its manufacturing processes by 95 percent or greater over its 1990 use, its products manufactured with MCF and/or CFC-113 are exempt from the labeling requirements. It must have achieved the above reduction either over the most recent calendar year, or for a 12-month period ending within 60 days of its certification to EPA.

Companies may submit certifications to EPA under this provision until May 15, 1994. Send certifications to: Labeling Manager, EPA, 6205-J, 401 M Street, SW, Washington, DC 20460. See final rule for further details on the certification and records, required.

Incidental Uses of Class I Substances
Labeling is not required for non-contact incidental uses of class I substances, including:

- A process in which a class I substance, such as a solvent, is used to clean or maintain manufacturing equipment, where the surface area being cleaned has no direct contact with the product.
- A process in which a class I substance is used in refrigerated equipment to keep food products cold; the refrigerant does not come into direct contact with the food products.

Labeling is not required for some *contact* incidental uses that are:

- A process in which a class I substance is used intermittently, not routinely, as part of the direct manufacturing process, such as spot-cleaning textiles, cleaning ink plates, or testing for leaks in a cooling system and condenser.
- A process in which there is an initial contact between the substance and the product, that occurs infrequently (typically as part of a maintenance process), and perhaps unintentionally. An example is the use of methyl chloroform as a spot remover in the textile industry; it is used infrequently and is not a routine part of the manufacturing process.

Labeling is required for uses that are not considered to be incidental. These include:

- Most mold release agents that are applied systematically throughout a manufacturing process.

- Defluxing of printed circuit boards during a continuous production process.

- Food processing, such as the manufacturing of some spices.

3) Products Containing Class I Substances and Containers of Class I or Class II Substances.

The distinction between containers and products containing containers of class I or class II substances or mixtures containing one of these substances must be labeled as of the effective date of the regulation. Products containing class I substances must also be labeled on May 15, 1993.

Products containing class II substances **will** be required to be labeled before January 2015 should the Administrator make the determination that substitutes are available for those products.

A *container* contains a class I substance if the substance must be transferred into another container or into another product in order to realize its intended use.

- Examples are a 5-gallon can of CFC-12, or an isotank of MCF, which would eventually be transferred to other vessels, such as refrigeration equipment or degreasing units, for their intended use.

A *product* contains a class I substance if the substance is used in the container or equipment without having to be transferred. Examples include some aerosols, solvents, adhesives, inks, coatings, and closed-cell foams.

Upon subsequent use of these products, however, a company would label its products manufactured with' (see example above in #2).

"Products containing" also include some air conditioning and refrigeration equipment. When they are installed in other products, such as automobiles, their labels would remain the same, because the refrigerant is intact at the point of purchase.

How to Label Containers of Recaptured Substances and Waste. If a company uses class I substances and captures those substances for incineration, it must label all containers of class I substance waste bound for incineration or waste containing trace amounts of class I substances.

Ozone-depleting substances bound for incineration and made into new products must be labeled to inform the service technician of the specific substances to be handled and processed. The new product containing ozone-depleting substances would require labeling.

Trace Quantities of Impurities Resulting From Inadvertent Production, Unreacted Feedstocks, and Process Agents

Inadvertent Production: EPA realizes there are circumstances in which an ozone-depleting substance is formed from a chemical reaction that takes place in a manufacture process, such as the formation of carbon tetrachloride in the chlorination of drinking water. Such production is unintentional, resulting in trace quantities of a class I substance remaining in the final product and therefore does not trigger the labeling requirements.

Process Agents: In addition, when manufacturers use an ozone-depleting substance as a feedstock or a process agent in their manufacturing processes and insignificant or trace quantities of the substance remain in the final product, the product is exempt from the labeling requirements for a "product containing."

For example, carbon tetrachloride is used as a catalyst in producing chlorinated rubber; the remaining trace amounts in the final product would not trigger the labeling requirements for a "product containing." This introduction of the substance is essential to the process and is neither consumed nor inadvertently produced, thus the final product would be labeled as a "product manufactured with."

In the case where a process agent is introduced, then removed from a product — such as in the case of many explosion suppressants — product would still require a label indicating it was manufactured with a class I substance,' unless the removed substance is subsequently transformed.

Treatment of the Use of Class I Substances for Repairs and Used Products
If a company sells solvent-cleaning products that do not contain ozone-depleting substances, and those products are used by manufacturers who may use ozone-depleting substances in their processes and then return the used solvent product for recycling, any contamination of those used non class I solvents resulting in trace amounts of class I substances in the recycled product would not trigger labeling.

If a company recycles solvents or other chemical products that contain class I substances necessary as part of their composition, it must label the new (recycled) product as a "product containing" a class I substance, since the substance is necessary to the manufacture of the product.

If a company sells used products, it is not required to re-label them, because they have already been introduced into interstate commerce.

If a company performs repairs or upgrades on products using class I substances, it is not required to label them; however, if it purchases components made with class I substances, those components should be labeled, but the label is not required to be passed through with the product. Products being upgraded, for example, would not require a new label on the final product, because they are not being introduced into interstate commerce.

4) Label Appearance and Placement

How the Labels Must Look
Format the labels so that they are in a *square or rectangular* area with or without a border. The word "WARNING" must be in *capital letters*. See rule for type size requirements.

The warning statement must be in *strong contrast against its background*. For example, black on white or red on white present strong contrasts; however, yellow on white or dark blue on green do not. The key is that the warning statement be *"clearly legible and conspicuous."*

The warning statement may be printed directly on a product or its outer packaging, or on alternative labeling; actual adhesive labels, although an option, are not required.

Where the Labels Should Appear

Principal Display Panel (PDP) or a Display Panel Area. Placing the warning statement on any of these display panels clearly meets the mandate that the statement be clearly legible and conspicuous.

Alternative Labels. These may be used, as long as the statement is clearly legible and conspicuous. Examples are hang tags, tape, cards, stickers, and other similar types of overlabeling.

Outer Packaging. The warning statement may be placed on the products outer packaging if the product is sold in its packaging, or if the consumer is able to read and understand the warning statement on such packaging at the time of purchase.

Supplemental Printed Materials. Placing the warning statement conspicuously in supplemental printed information that accompanies the product or container such as invoices, bills of lading, package inserts, and Material Safety Data Sheets (MSDS) at the time of purchase meets the requirements as long as the purchaser can read the warning statement upon purchase.

Promotional Materials. For products purchased through telephone or mail orders, print the warning statement in a conspicuous place in sales promotional literature, journals, newspapers, or displays so the warning statement is available before the time of purchase. A company could include an insert in such printed material bearing the warning statement. Another option would be the use of supplemental printed materials that accompany the product at the time of delivery. With this option, the consumer must be able to return the product if, at the time of delivery or payment, they choose not to make the purchase based on the warning statement.

Products that are labeled in any one of the above methods do not require additional labeling.

5) Treatment of Products Manufactured for Export

Products manufactured for export are not required to be labeled, but there must be sound evidence that such products are intended for export. This could include:

- Clear identification of an export area in a warehouse.

- Destination papers, shipping papers, or other documentation indicating that the products are intended for export. This information must be readily available on site upon request by EPA.

6) Treatment of Products Manufactured for Import

The importer will be held liable for all products subject to the labeling requirements imported into the United States. These products are introduced into interstate commerce at the site of U.S. Customs clearance.

Importers must have a reasonable belief that products introduced into interstate commerce are accurately labeled. In order to have a reasonable belief, importers may investigate at least one step back into the manufacturing process or develop a contractual agreement with its supplier indicating whether the products have been made with ozone-depleting substances. An MSDS would be another option for establishing a reasonable belief.

An importer planning to incorporate its imported product made with class I substances into a new product must label the import; however, it is not required to label its final product if no class I substance is used in the manufacture of the final product. If the product contains a class I substance, such as an air conditioner to be installed into an automobile, the importer must label the final product (the vehicle) as a "product containing" upon its introduction into interstate commerce.

7) Labeling of Packaging Materials

Manufacturing of packaging Materials Made With Class I Substances Must Label
If a company makes packaging materials using class I substances, it must label its products as 'manufactured with a class I substance.' The company's customer, however, is not required to pass the label on the packaging materials through with its product. Examples of these products include some corrugated packaging or open-cell foam-blown materials.

Manufacturers of Packaging Materials Used to Package Other products Made by That Company
If a company makes its own packaging materials using a class I substance and it also makes a product using no ozone-depleting substances to be sold in those packaging materials, its final package must be labeled based on its use of class I substances in the packaging materials.

For example, a candy company may use a class I adhesive to affix a wrapper. The final product would be labeled "manufactured with."

8) Products and Processes Under Research & Development

The use of class I or class II substances in the research and development of a product or process does not require labeling, since neither one has been introduced into interstate commerce. Upon a new product's introduction into interstate commerce, labeling would be required.

9) EPA Petition Processes

A manufacturer may petition EPA to add a class II substance or process using such a substance to the labeling regulation if EPA determines that substitute products or processes: do not rely on class II substances; reduce the overall risk to human health and the environment and; are currently or potentially available.

For products made with class I substances, a manufacturer may petition EPA to temporarily exempt a product or process using a class I substance from the labeling requirements if EPA determines that no substitute products or processes exist that: do not rely on class I substances; reduce the overall risk to human health and the environment and are currently or potentially available.

10) Introduction Into Interstate Commerce

There are three entry points into interstate commerce for purposes of labeling requirements

- Site of U.S. Customs (**Customs**) clearance. Labeling may occur: (1) at the foreign production facility per agreement between manufacturer and importer, while in transit, or at another location before U.S. Customs inspects goods; (2) in supplemental printed material prior to the products entering the location in which customs inspection occurs.

- Introduction into manufacturer's, distributor's, wholesaler's or retailer's warehouse. Labeling may occur at manufacturer's production facility, or upon entry into manufacturer's warehouse.

  Distributors, wholesalers and retailers must pass labeling information through to the customer. Labeling information received from a manufacturer must remain with products in the warehouse until the distributor, wholesaler, or retailer distributes or sells them. If the labeled products are repackaged or require new labels, such labeling must be secured prior to the release of the products from the warehouse.

For example, a distributor purchases a bulk shipment of nuts and bolts that are labeled on supplemental printed material as products manufactured with CFCs. The supplemental printed material must remain with the products in inventory until the distributor is ready to sell them. Prior to the sale, the distributor may repackage the products into smaller boxes and label the individual boxes. The labeling must be secured prior to the release of the products.

- Release from manufacturer's production facility where manufacturer has no warehouse. Labeling may occur during production on production line, at end of production line, or any time prior to release of products from production facility.

Labeling Charged Containers
A container charged with a class I substance must be labeled either when it leaves the place of charging activity, when it enters storage for further sale, or when it enters a site of U.S. customs clearance. Containers kept within a facility for a company's own manufacturing purposes need not be labeled because they are not being introduced into interstate commerce.

Recharging "Products Containing"
A halon manufacturer or distributor refills a fire extinguisher for use in a customer's facility. The container discharging the halon would be labeled as a "container containing"; however, the fire extinguisher, already purchased, would not require additional labeling. Fire extinguishers would be labeled as "products containing" when they are sold. Subsequent labeling upon refilling activity would not be required.

Servicing of "products containing" such as degreasers, fire extinguishers, and air conditioners would not require re-labeling by distributors or manufacturers.

EPA cannot hold servicers liable for removal of labels by customers. Furthermore, EPA exempts the use of ozone - depleting substances from the labeling requirements for repairs.

EFFECTIVE DATES

- Regulation effective May 15, 1993, pursuant to section 611.
- Because this rule was issued only three months prior to the date compliance is required, there may be a need in some cases for additional time to meet the requirements. At this time, therefore, it is the Agency's policy not to take enforcement actions for matters occurring during 9 months following the date of publication (February 11, 1993).
- EPA expects the regulated community to meet the statutory deadline of May 15, 1993.
- The grandfather provision, discussed above, only applies to those products made prior to May 15, 1993, despite any enforcement discretion afforded.

FOR ADDITIONAL INFORMATION

Contact the Stratospheric Ozone Information Hotline at 800 296-1996, Monday-Friday, between the hours of 10:00 a.m.-4:00 p.m. (Eastern). International callers must dial 202 783-1100.

To receive copies of the final rule and any follow-up activity regarding the labeling rule, such as notices on petitions and policy on a destruction exemption from the labeling requirements, see the *Federal Register* in a local university or in government libraries, or call the hotline.

State And Territorial Air Pollution Control Agencies

Alabama Dept. of Environmental Management
Air Division
1751 Cong. Dickenson Drive
Montgomery, AL 36130
(205) 271-7861

Alaska Dept. of Environmental Conservation
Air Quality Management Section
P. 0. Box 0
Juneau, AK 99811-1800
(907) 465-5100

American Samoa
Environmental Quality Commission
Governor's Office
Pago Pago, AM. Samoa 96799
Oll-(684) 633-4116

Arizona Dept. of Environmental Quality
P.0, Box 600
Phoenix, AZ 85001-0600
(602) 257-2308

Arkansas Dept. of Pollution Control and
Ecology
Air Division
9001 National Drive, P.O. Box 9583
Little Rock, AR 72209
(501) 562-7444

California Air Resources Board
Secretary of Environmental Affairs
P. 0. Box 2815
Sacramento, CA 95812
(916) 445-4383

Colorado Dept. of Health
Air Pollution Control Division
4210 E 11th Avenue
Denver, CO 80220
(303) 331-8500

Connecticut Dept. of Environmental Protection
Bureau of Air Management
165 Capitol Avenue
Hartford, CT 06106
(203) 566-2506

Delaware Dept. of Nat. Resources
and Environmental Control
Division of Air and Waste Mgmt.
89 Kings Highway, P.O. Box 1401
Dover, DE 19903
(302) 739-4791

District of Columbia Dept. Cons. and Reg. Affairs
Air Quality Control and Monitoring Branch
2100 Martin Luther King Ave. SE
Washington DC 20020
(202) 404-1120

Florida Dept. of Environmental Regulation
Air Resources Mgmt.
2600 Blair Stone Road
Tallahassee, FL 32399-2400
(904) 488-1344

Georgia Dept. of Natural Resources
Air Resources Branch
205 Butler Street, SE
Atlanta, GA 30344
(404) 656-6900

Guam Environmental Protection Agency
Complex Unit D-107
130 Rojas Street
Harmon, Guam 96911
Oll-(671) 646-8863

Hawaii State Dept. of Health
Laboratories Div. Air Surveillance
Analysis Branch
1270 Queen Emma St., Suite 900
Honolulu, HI 96813
(808) 586-4019

Idaho Division of Environmental Quality
Air Quality Bureau
1410 North Hilton
Boise, ID 83706

Illinois Environmental Protection Agency
Division of Air Pollution Control
2200 Churchill Road, P.O. Box 19276
Springfield, IL 62794-9276
(217) 782-7326

Indiana Dept. of Environmental Management
Office of Air Management
105 S. Meridian Street, P.0, Box 6015
Indianapolis.. IN 46206-6015
(317) 232-8384

Iowa Dept. of Natural Resources
Air Quality Section
Henry A. Wallace Building, 900 E. Grand St.
Des Moines, IA 50319 - (515) 281-8852

Kansas Dept. of Health and Environment

Bureau of Air and Waste Management
Forbes Field, Building 740
Topeka, KS 66620
(913) 296-1593

Kentucky Dept. for Environmental Protection
Division for Air Quality
316 St. Clair Mall
Frankfort, KY 40601
(502) 564-3382

Louisiana Dept. of Environmental Quality
Office of Air Quality and Radiation Protection
Air Quality Division, P.O. Box 82135
Baton Rouge, LA 70884-2135
(504) 765-0110

Maine Dept. of Environmental Protection
Bureau of Air Quality Control
State House, Station 17
Augusta, ME 04333
(207) 289-2437

Maryland Dept. of the Environment
2500 Broening Highway
Baltimore, MD 21224
(301) 631-3255

Massachusetts Dept. of Environmental
Protection
Division of Air Quality Control
One Winter St., 5th Floor
Boston, MA 02108
(617) 292-5593

Michigan Dept. of Natural Resources
Air Quality Division
P.O. Box 30028
Lansing, MI 48909
(517) 373-7023

Mississippi Dept. of Environmental Quality
Air Division, Office of Pollution Control
P.O. Box 10385
Jackson, MS 39299
(601) 961-5171

Minnesota Pollution Control Agency
Air Quality Division
520 Lafayette Road
Saint Paul, MN 55155
(612) 296-7331

Missouri Dept. of Natural Resources
Division of Env. Quality, Air Pollution Control
P.O. Box 176

Jefferson City, MO 65102
(314) 751-4817

Montana Dept. of Health and Environmental
Science
Air Quality Bureau
Cogswell Building, Room A116
Helena, MT 59620
(406) 444-3454

New Mexico Environmental Department
Air Quality Division, Env. Prot. Div.
P. 0. Box 26110
Santa Fe, NM 87502
(505) 827-0070

New York Dept. of Environmental Conservation
Division of Air Resources
50 Wolf Road
Albany, NY 12223-3250

Nebraska Dept. of Environmental Control
Air Quality Division
P, 0. Box 99922
Lincoln, NB 68509-8922
(402) 471-2189

Nevada Division of Environmental Protection
Bureau of Air Quality
123 West Nye Lane
Carson City, NV 89710
(702) 687-5065

New Jersey Div. of Environmental Quality
401 East State Street
Trenton, NJ 08625
(609) 292-6710

New Jersey Dept. of Environmental Protection
401 East State St.
Trenton, NJ 08625
(301) 631-3255

New Hampshire
Air Resources Division
64 N. Main Street, Box 2033
Concord, NH 03301
(603) 271-1370

North Dakota State Dept. of Health
Division of Environmental Engineering
1200 Missouri Avenue
Bismarck, ND 58502-5520
(701) 221-5188

North Carolina Dept. of Environmental Health and

Natural Resources, Air Quality Section
P. 0. Box 27687
Raleigh, NC 27611-7687
(919) 733-3340

Ohio Environmental Protection Agency
Division of Air Pollution Control
1800 Watermark Drive
Columbus, OH 43266-0149
(614) 644-2270

Oklahoma State Dept. of Health
Air Quality Service
1000 Northeast 10th Street
P.O. Box 33551
Oklahoma City, OK 73152
(405) 271-5220

Oregon Dept of Environmental Quality
Air Quality Control Division
811 SW 6th Avenue, 11th Fl.
Portland, OR 97204
(503) 229-5287

Pennsylvania Dept. of Environmental Resources
Bureau of Air Quality Control
101 South Second St.
Harrisburg, PA 17105-2357
(717) 787-9702

Puerto Rico Environmental Quality Board
Edificio Banco National Plaza
431 Ave Ponce DeLeon
Hato Rey, PR 00917
(909) 767-8071

Rhode Island Dept. of Environmental Mgmt.
Division of Air and Hazardous Materials
291 Promenade St.
Providence, RI 02908-5767
(401) 2?7-2908

South Dakota Dept. of Environmental and
Nat. Resources, Point Source Program
523 East Capitol Avenue
Pierre, SD 57501
(605) 773-3153

South Carolina Dept. of Health
Bureau of Air Quality Control
2600 Bull Street
Columbia, SC 29201
(803) 734-4750
(512) 908-1000

Texas Air Control Board
12124 Park 35 Circle
Austin, TX 78753
(208) 334-5898

Tennessee Dept. of Environment and
Conservation
Division of Air Pollution Control
701 Broadway
Nashville, TN 37243-1531
(615)741-3931

Utah Dept. of Environmental Quality
Division of Air Quality
1950 West North Temple
Salt Lake City, UT 84114-4820
(801) 536-4000

Vermont Agency of Natural Resources
Air Pollution Control Division
103 S. Main St. Building 3 South
Waterbury, VT 05676
(802) 244-8731

Virgin Islands Dept. Div. of Environmental Protection
Watergut Homes 1118 Christiansted
St. Croix, VI 00820-5065
(809) 773-0565

Virginia Department of Air Pollution Control
P.O. Box 10089
Richmond, VA 23240
(804) 786-2379

Washington State
Department of Ecology
P. 0. Box 47600
Olympia, WA 98504-7600
(206) 459-6632

West Virginia Air Pollution Control Commission
1558 Washington St. East
Charleston, WV 25311
(304) 348-2275

Wisconsin Dept. of Natural Resources
Bureau of Air Management
Box 7921
Madison, WI 53707
(609) 266-7719

Wyoming Dept. of Environmental Quality
Air Quality Division
122 W, 25th St.
Cheyenne, WY 82002
(307) 777-7391

SPORLAN PRESSURE–TEMPERATURE CHART

PSIG	Green	Purple	Yellow	Blue	White
	Temperature, °F				
	REFRIGERANT - (Sporlan Code)				
	22 (V)	502 (R)	12 (F)	134a (J)	717 (A)
5*	-48	-57	-29	-22	-34
4*	-47	-55	-28	-21	-33
3*	-45	-54	-26	-19	-32
2*	-44	-52	-25	-18	-30
1*	-43	-51	-23	-16	-29
0	-41	-50	-22	-15	-28
1	-39	-47	-19	-12	-26
2	-37	-45	-16	-10	-23
3	-34	-42	-14	-8	-21
4	-32	-40	-11	-5	-19
5	-30	-38	-9	-3	-17
6	-28	-36	-7	-1	-15
7	-26	-34	-4	1	-13
8	-24	-32	-2	3	-12
9	-22	-30	0	5	-10
10	-20	-29	2	7	-8
11	-19	-27	4	8	-7
12	-17	-25	5	10	-5
13	-15	-24	7	12	-4
14	-14	-22	9	13	-2
15	-12	-20	11	15	-1
16	-11	-19	12	16	1
17	-9	-18	14	18	2
18	-8	-16	15	19	3
19	-7	-15	17	21	4
20	-5	-13	18	22	6
21	-4	-12	20	24	7
22	-3	-11	21	25	8
23	-1	-9	23	26	9
24	0	-8	24	27	11
25	1	-7	25	29	12
26	2	-6	27	30	13
27	4	-5	28	31	14
28	5	-3	29	32	15
29	6	-2	31	33	16
30	7	-1	32	35	17
31	8	0	33	36	18
32	9	1	34	37	19
33	10	2	35	38	19
34	11	3	37	39	20
35	12	4	38	40	21
36	13	5	39	41	22
37	14	6	40	42	23
38	15	7	41	43	24
39	16	8	42	44	25
40	17	9	43	45	26
42	19	11	45	47	28
44	21	13	47	49	29
46	23	15	49	51	31
48	24	16	51	52	32
50	26	18	53	54	34
52	28	20	55	56	35
54	29	21	57	57	37
56	31	23	58	59	38
58	32	24	60	60	40
60	34	26	62	62	41
62	35	27	64	64	42
64	37	29	65	65	44
66	38	30	67	66	45
68	40	32	68	68	46
70	41	33	70	69	47
72	42	34	71	71	49
74	44	36	73	72	50
76	45	37	74	73	51
78	46	38	76	75	52
80	48	40	77	76	53
85	51	43	81	79	56
90	54	46	84	82	58
95	56	49	87	85	61
100	59	51	90	88	63
105	62	54	93	90	66
110	64	57	96	93	68
115	67	59	99	96	70
120	69	62	102	98	73
125	72	64	104	100	75
130	74	67	107	103	77
135	76	69	109	105	79
140	78	71	112	107	81
145	81	73	114	109	82
150	83	75	117	112	84
155	85	77	119	114	86
160	87	80	121	116	88
165	89	82	123	118	90
170	91	83	126	120	91
175	92	85	128	122	93
180	94	87	130	123	95
185	96	89	132	125	96
190	98	91	134	127	98
195	100	93	136	129	99
200	101	95	138	131	101
205	103	96	140	132	102
210	105	98	142	134	104
220	108	101	145	137	107
230	111	105	149	140	109
240	114	108	152	143	112
250	117	111	156	146	115
260	120	114	159	149	117
275	124	118	163	153	121
290	128	122	168	157	124
305	132	126	172	161	128
320	136	130	177	165	131
335	139	133	181	169	134
350	143	137	185	172	137
365	146	140	188	176	140

* Inches mercury below one atmosphere

PRESSURE–TEMPERATURE CHART

Temperature, °F

PSIG	Pink	Orange	Sand	Teal	Green	Tan
	REFRIGERANT - (Sporlan Code)					
	MP39 (X) or 401A (X)	HP62 (S) or 404A (S)	HP80 (L) or 402A (S)	AZ-50(P) or 507(P)	124 (Q)	125
5*	-23	-57	-59	-59	3	-63
4*	-22	-56	-58	-57	4	-61
3*	-20	-54	-56	-56	6	-60
2*	-19	-53	-55	-55	7	-58
1*	-17	-52	-54	-53	9	-57
0	-16	-51	-53	-52	10	-56
1	-13	-48	-50	-50	13	-53
2	-11	-46	-48	-47	16	-51
3	-9	-43	-45	-45	18	-49
4	-6	-41	-43	-43	21	-46
5	-4	-39	-41	-41	23	-44
6	-2	-37	-39	-39	26	-42
7	0	-35	-37	-37	28	-40
8	2	-33	-36	-35	30	-39
9	4	-32	-34	-34	32	-37
10	6	-30	-32	-32	34	-35
11	8	-28	-30	-30	36	-33
12	9	-27	-29	-29	38	-32
13	11	-25	-27	-27	40	-30
14	13	-23	-26	-25	41	-29
15	14	-22	-24	-24	43	-27
16	16	-20	-23	-23	45	-26
17	17	-19	-21	-21	46	-24
18	18	-18	-20	-20	48	-23
19	20	-16	-19	-18	49	-22
20	21	-15	-17	-17	51	-20
21	23	-14	-16	-16	52	-19
22	24	-12	-15	-15	54	-18
23	25	-11	-14	-14	55	-16
24	27	-10	-12	-12	57	-15
25	28	-9	-11	-11	58	-14
26	29	-8	-10	-10	59	-13
27	30	-6	-9	-9	61	-12
28	32	-5	-8	-8	62	-11
29	33	-4	-7	-7	63	-10
30	34	-3	-6	-5	65	-8
31	35	-2	-5	-4	66	-7
32	36	-1	-4	-3	67	-6
33	37	0	-3	-2	68	-5
34	38	1	-1	-1	69	-4
35	39	2	0	0	71	-3
36	40 (30)	3	0	1	72	-2
37	41 (31)	4	1	2	73	-1
38	43 (32)	5	2	3	74	0
39	44 (33)	6	3	4	75	0
40	45 (34)	7	4	5	76	1
42	46 (36)	8	6	6	78	3
44	48 (38)	10	8	8	80	5
46	50 (40)	12	10	10	82	7
48	(42)	14	11		84	8
50	44	16 (13)	13	13	86	10
52	45	17 (14)	15	15	88	11
54	47	19 (16)	16	16	90	13
56	49	20 (18)	18	18	91	15
58	50	22 (19)	19	19	93	16
60	52	23 (20)	21	21	95	17
62	53	25 (22)	22	22	97	19
64	55	26 (23)	24	24	98	20
66	56	27 (25)	25	25	100	22
68	58	29 (26)	27	27	101	23
70	59	30 (27)	29	28	103	24
72	61	32 (29)	31	29	104	26
74	62	33 (30)	32	30	106	27
76	64	34 (31)	33	32	107	28
78	65	35 (32)	34	33	109	29
80	66	37 (36)	34 (31)	34	110	31
85	69	40 (39)	37 (34)	37	114	33
90	73	42 (42)	40 (37)	40	117	36
95	76	45 (44)	42 (40)	43	120	39
100	78	48 (47)	45 (43)	46	123	42
105	81	50	48 (45)	48	126	44
110	84	52	50 (48)	51	129	47
115	87	55	50	53	132	49
120	89	57 (59)	53	56	135	51
125	92	59	55	58	138	54
130	94	62	57 (60)	60	140	56
135	96	64	60 (62)	62	143	58
140	99	66	62 (64)	64	145	60
145	101	68	64 (67)	67	148	62
150	103	70	66 (69)	69	150	64
155	105	72	68	71	152	66
160	108	74 (76)	70 (72)	73	154	68
165	110	76	72 (74)	74	157	70
170	112	78	74 (76)	76	159	72
175	114	80	75 (78)	78	161	73
180	116	82	77 (80)	80	163	75
185	117	83	79 (82)	82	165	77
190	119	85	81 (83)	83	167	79
195	121	87	82 (85)	85	169	80
200	123	88	84 (87)	87	171	82
205	125	90	86 (88)	88	173	83
210	127	92	87 (90)	90	175	85
220	130	95	91 (93)	93	178	88
230	133	98	94 (96)	96	182	91
240	136	101	97 (99)	99	185	94
250	140	104	99 (102)	102	188	97
260	143	107	102 (105)	105	192	99
275	147	111	106 (109)	109	196	103
290	151	115	110 (112)	112	201	107
305	155	118	114 (116)	116	205	111
320	159	122	118 (120)	120	209	114
335	163	126	121 (123)	123	213	118
350	167	129	125 (126)	126	217	121
365	170	132	128 (129)	129	221	124

Vertical labels within the blend columns indicate BUBBLE POINT and DEW POINT regions.

* Inches mercury below one atmosphere

P-H DIAGRAM – BLENDS

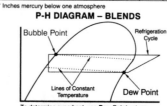

Bubble Point — Refrigeration Cycle — Lines of Constant Temperature — Dew Point

To determine superheat, use **Dew Point** values.
To determine subcooling, use **Bubble Point** values.

What's your Superheat?

EXAMPLE
REFRIGERANT —22—
TEMPERATURE HERE READS 52°
OBTAIN SUCTION PRESSURE ......68 PSIG (at bulb)
CONVERTED TO TEMP 40°
12° SUPERHEAT

SPORLAN VALVE COMPANY   ST. LOUIS, MO 63143
FORM 1-100-194   COPYRIGHT 1994 BY SPORLAN VALVE COMPANY, ST. LOUIS, MO 63143   PRINTED IN U.S.A.

APPENDIX VII

TRAINING PLANS

Training programs allow you and your company to efficiently prepare for EPA compliance. Incorporating EPA recommended service procedures into your training plan is an effective and low cost way to reenforce recommended practices. Written plans document that company technicians have received sufficient training and know the proper procedures and how to apply them. All shops, no matter how small, will benefit from a well developed training plan. A little planning can return BIG dividends.

Most shop training plans focus on purely technical procedures. A truly comprehensive training plan incorporates more than just technical needs. Chapter 10 touches on behavior that changes attitudes. Those items should be added to your training plan. Incorporate section 608, 609 and 611 (labeling) requirements into the plan and specify each section that is quoted, i.e. (82.146) for the 611 labeling section. Doing so, makes Clean Air Act compliance second nature and it also reinforces the use of EPA recommended service procedures. The same should be done if there are state or local requirements. I see a blending of management, technical and safety practices incorporated into training plans as a key factor in developing a business that progressively complies with environmental issues.

A good training plan can be developed through basic questioning techniques. Ask yourself and the group what must be accomplished to comply with the Clean Air Act. Your answers will provide the *nuts-and-bolts* of the plan. Review the brainstorming techniques on page 74 prior to developing your plan. The following questions are presented to help you focus on training issues.

TYPICAL TRAINING QUESTIONS
(LIST VIII - 1)

1.   What do you want to accomplish with this training? What level or depth of training is needed? i.e.,

☐   Correct use of shop recovery equipment.
☐   Documentation of refrigerant saved to holding containers.
☐

2.   Do you have a union shop? If you are a union shop, work with the shop steward to develop a compliance partnership between management and labor.

3.   Who needs this training?

☐   Make a list of all employee's training needs.

4.   Will training be done in the shop by a knowledgeable technician or by a factory representative or private contractor?

☐   If training is provided from a contractor obtain certificates of training even if you have to provide the certificates. Add this requirement to the trainers contract.

5.   What test equipment & instruments are needed for the training?

☐   Recovery/recycling equipment.
☐   List all needed equipment.

6.   What preparation is needed before the training and who needs to do it?

7.   What will be the expenses or estimates of cost to support your training plan?

☐   On duty study time for technicians. (hours) ____
☐   Cost of study materials.
✎   Cost of this study guide times the number of technicians. $____ (# of Techs ____ x $29.95)
☐   Training course costs (if used). $____
☐   Do you need new equipment to support the training?
☐   EPA Certification Exam Cost. $____
☐   Other costs. (list)

8.   List the desired training results that are indicators of successful training accomplishment.

❑   A practice written test.
❑   A practice performance examination.
✎   You can use the closed book practice exam in Appendix III to evaluate technician comprehension.
❑   Passing the EPA Certification exam.

9.   How will you give recognition to those who successfully complete the training?

❑   Training institution issues training certificate.
✎   Insure training institution is EPA certified.
❑   Use a generic training certificate.
✎   Ideal for technicians that took the EPA exam after studying this EPA study guide.

10.   How will you schedule training?

❑   When, how often, and where.
✎   Is there a federal, state or local time frame for completion? Do you need to establish your own time requirements?

11.   How do you want to track and record the training? If the training is required by federal, state or local governments what are their records requirements?

❑   Establish a separate training file for each technician.
✎   Include copy of training plan, training certificates, record of hours/time provided for EPA certification and other training.

12.   Where will the training be conducted?

❑   Atmosphere   ❑   Lighting   ❑   Desk space

13.   What other materials do you need to improve the training?

❑   Order a copy of *Air Conditioning & Refrigeration Technician's EPA Certification Guide* for each supervisor, manager and technician. *#___* needed. (Order form on page 191 in the back of this book)
❑   List all materials needed.

14.   Do you have the necessary safety equipment for conducting the training?

❑   List required safety equipment.

15.   How will you assess the training?

❑   UDevelop a training critique sheet.

16.   What level or depth of training is required of the course and for the technicians?

17.   What, if any, management goals do you want to include and where do you want them in the plan?

❑   Full compliance with the Clean Air Act.
❑   A safe working environment.
❑   All technicians to pass the EPA examination.

DEVELOPING THE TRAINING PLAN
(LIST VIII-2)

The previous questions were developed to prepare shops for the training process. The following list will help you focus on specific technical training issues.

1. What is the equipment, procedure or process to be trained on?

❑   Recovery/Reclaiming process
❑   Prevention of venting
❑   Leak detection and repair
❑   Labeling/storage/disposal
❑   _____
❑   _____

2. List the major tasks or objectives the technician must do.

❑   Shut down system
❑   Isolate compressor
❑   Recovery equipment hookup/use
❑   _____
❑   _____

3. Break the training tasks into parts, elements or groups.

4. If possible further reduce to sub-parts, elements or groups.

5. List each step to be done in 2,3,4.

6. If appropriate, state what completing each part or sub-part accomplished.

✎ Example: The compressor suction and pressure lines are closed, compressor electrical supply is disconnected, tagged and locked out, and verified there is no voltage to the appliance. Now we can remove the refrigerant charge from the compressor.

7. After completing 2,3,4 and 5, flag any key procedure that specified results were not obtained and identify what should be done.

> ✎ Example: a. Start over, b. follow all steps in reverse order back to the starting point, c. backup one step and try again.

8. After completing 2 thorough 7, flag those steps that are safety concerns and specify what must be done if there is an incident BEFORE doing the step. Use a star or asterisk (*) to identify safety concerns.

9. Are there critical steps that everyone must follow to achieve the training objective? If yes, mark critical steps with a star.

10. Give each participant a handout outlining the training sessions activities.

> ❑ Provide copies of training plans or course outlines to all trainees.

SAFETY — SAFETY — SAFETY

SAFETY BRIEFINGS must be an integral part of all training. I recommend using a checklist for safety briefings. Have every participant sign and place the date of the briefing next to their signature at the bottom of the checklist before the training starts.

SAFETY CHECK LIST
(LIST VIII-3)

❑ Show trainees where the electrical disconnect switch is located for the equipment that will be worked on. Insure that all CIRCUIT BREAKERS are properly marked.
❑ Review emergency procedures.

> ✎ Identify who in the group has CPR training in the event of electrical shock.

> ✎ Post emergency numbers for ambulance services, emergency medical service providers, local hospitals, fire departments, police, and hazmat coordinators on the wall next to the shop phone. Brief everyone on the phone locations.

> ✎ Point out the location of fire extinguishers and first aid kits.

> ✎ Add to this list for your particular circumstances.

❑   Identify mechanical rotating or moving parts that can cause injury or death.

   ✎ Post caution and hazard signs where appropriate.

   ✎ Show trainees where quick disconnect switches are.

❑ Insure that trainees have the proper personal protective equipment required for the training.

❑ Review RED CROSS posters for administering CPR and other medical emergencies.

❑ Review hazards associated with refrigerants and other compressed gasses.

❑ Cover hazards associated with brazing torches and flammable materials.

❑ Review Material Safety Data Sheets (MSDS) for all chemicals used or encountered during training.

   ✎ Identify the shop location for the MSDS binder and explain how to read the MSDS sheets in the event of contamination.

❑ Point out any other safety equipment available such as eyewash stations, shower area or safety boards.

❑ In case of fire or a gas leak point out the exits and ensure they are not locked or blocked.

❑ De-energized equipment can KILL. Always measure circuits with an AC and DC meter to ensure capacitors are discharged before touching.

❑ Repeat the safety briefing again after lunch breaks.

   Place this list in the front of the written training plan. Someone may use the plan for refresher training on their own. If an item does not apply, state that in the safety briefing. Following these habits will help maintain a safe working environment.

A SAMPLE TRAINING PLAN

* = Safety item.

1. Train all technicians on Digital Multimeter.
   ✎ Provide plan overview.

2. Major tasks: (overview)

Cover safety precautions

AC scale

DC scale

OHMS/Resistance scale

Current scale

Servicing digital multimeters

3. Major tasks: (procedures)

A. Give safety briefing

1 Review Safety Check List (VIII-3)
2 Cover safety precautions in manufacturer's book
3 Inspect test leads for cracks or exposed wire
4 Dangers of overloading instrument
5 Before touching a "DEAD" circuit, test with AC and DC scales
6 Dangers of capacitors holding a charge; describe proper discharge method

B. AC scale

1 *Discuss why to always set meter to highest scale
2 *Show proper way to take measurements
3 Demonstrate over ranging
4 Demonstrate under ranging
5 Set meter to off and have technician take measurement
6 *Observe technician setting scale for measurement
7 *Observe proper use of test probes to take measurement
8 Check to see if the technician read the display correctly

C. DC scale

1 *Set to highest scale
2 *Show proper way to measure DC voltages, probe polarity
3 Discuss measurement term, point A with respect to B
4 Show example of point A with respect to B
5 Show how a change in point of reference changes reading
6 Show how AC voltage can be on DC supply and how to measure
7 Set meter to off and have technician take measurement
8 *Observe technician setting scale for measurement
9 *Observe for proper use of test probes to take measurement.

D. Ohms/Resistance scale

1 *Ensure no voltages or charged capacitors in measurement path
2 Demonstrate lowest scale reading
3 Demonstrate highest scale reading
4 *Discuss and show differences of digital and analog meter testing of capacitors
5 Show that high scale settings can measure wire insulation giving a false indication of leakage.

This partial digital meter training plan is presented to show you a typical plan's structure and it can be used as a guide for developing your plans. The plan follows steps 1 thru 9 with a safety briefing at the start. Asterisks are placed in front of steps that could result in damage to the instrument or injury to personnel demonstrating proficiency.

TRAINERS QUESTIONING TECHNIQUES

To promote discussion or determine understanding, use questions that begin with WHO, WHAT, WHERE, WHEN, WHY AND HOW.

EXAMPLES

Why isolate the suction and pressure lines to the compressor?
When do you need to reverse your meter leads?
What circumstances warrant having recovery equipment available?

If the discussion stops and you want to start it up, use the following techniques.

EXAMPLES

What have we missed that is important to this task?
Are there alternative methods or procedures that we can use that do not violate EPA practices or rules?

Put the word suppose or what would happen at the beginning of a sentence to get people thinking.

EXAMPLES

Suppose I remove the boost capacitor while the system is running, what would happen?

What would happen if your leads were reversed while measuring a DC voltage?

INDEX

CAREER RESOURCES

AIR CONDITIONING & REFRIGERATION TECHNICIAN's

EPA CERTIFICATION GUIDE

Getting Certified, Understanding the Rules, & Preparing for EPA Inspections

Learn from an expert. The author, James Preston — certified universal technician — passed the **Type I, II, III, and Universal** exams on his first try using the techniques presented in this book.

All air conditioning and refrigeration technicians now require certification by the EPA.

This certification and training guide teaches service technicians, those planning to enter the field, and business owners about the 1990 Clean Air Act's refrigeration recycling rules and prepares them for the certification test and on-site EPA inspections.

Don't be left out in the cold! Joe Cronley, Publisher of **REFRIGERATION News** said, *"Many technicians are going to avoid getting certified. Unfortunately, their employers can't keep them."* Think again about waiting to be "grandfathered" in. This guide prepares technicians, students, and businesses for   Clean Air Act implementation and includes:

- Certification Test Preparation
- Training Plan Development
- Freon Recovery/Recycling
- Labeling Requirements
- Complete References

- Helpful Checklists
- EPA Inspection Guidance
- Closed Book Practice Exams
- EPA Record Keeping Requirements
- The History Behind Ozone Depletion

"...Since the announcement of EPA regs mandating technician testing last fall, there has been a tremendous demand for a guidebook which will 'do it all': get your employees certified, get your equipment in order, and make sure you're OK with the EPA... This book may be your best resource you have, short of a paid consultant. A valuable book."

— **Joe Cronley, Publisher**
REFRIGERATION News

- ISBN: 0-943641-10-1
- 192 pages, 8½" x 11", $29.95
- LC # 94-071513, Publication Date: 7/94
- Illustrated/Index/Appendices

PLEASE TURN PAGE

THE BOOK OF U.S. GOVERNMENT JOBS, 5th Ed., by Dennis Damp

$15.95, 1994, 224 pages

Where They Are, What's Available, & How to Get One. The all new 4th edition is an easy to follow, step-by-step guide to high-paying jobs with the U.S. government. Filled with everything you need to know about obtaining a federal government job, this book guides you through Uncle Sam's unique hiring world..

"A comprehensive how-to-guide. Recommended ... A worthy, affordable addition to career collections."   **— LIBRARY JOURNAL**

" ... A TREASURE TROVE OF INFORMATION."   **— BOOKWATCH**

HEALTH CARE JOB EXPLOSION! Career In The 90's by Dennis V. Damp.

$14.95, 1993, 384 pages

Audiologist
Chiropractor
Clinical Laboratory Technologist
Dental Hygienist
Dental Laboratory Technician
Dentist
Dietitian
Dispensing Optician
EEG & EKG Technologist
Emergency Medical Technician
Homemaker-Home Health Aides
Human Service Worker
Licensed Practical Nurse
Medical Assistants
Medical Record Technician
Nuclear Medicine Technologist
Nursing Aide
Nutritionist
Occupational Therapist
Ophthalmic Lab Technician
Optometrist
Pharmacist
Psychiatric Aide
Physical Therapist
Physician
Physician Assistant
Radiologic Technologist
Recreational Therapist
Registered Nurse
Respiratory Therapists
Social Worker
Speech-Language Pathologist
Surgical Technologist
Veterinarian

The health care job market is **EXPLODING**. Currently, seven out of every one hundred Americans work in health services. By the year 2005 that figure will increase to nine out of every one hundred, that's **3,900,000 NEW JOBS!**

Health Care Job Explosion! by Dennis V. Damp is a comprehensive **career guide** and **job finder** that steers readers to where they can actually find job openings; periodicals with job ads, placement services, directories, associations, job fairs, and job hotlines. All major health care groups are explored including the nature of work for each occupation, describing:

- Typical working conditions
- Training/advancement potential
- Job outlook and earnings
- Employment opportunities
- Necessary qualifications
- Related occupations

PLUS more than 1,000 verified job resources

"...Solid job-seeking hints. Info is plentiful, comprehensive, & thorough. "
— AMERICAN BOOKSELLER
"...This book will be a boon to those seeking jobs. Recommended. "
— LIBRARY JOURNAL REVIEW
"...Excellent info for job-conscious collegians."
— THE BLACK COLLEGIAN

PLEASE TURN PAGE

ORDERING INFORMATION

Use the order form that follows to purchase the titles that interest you. Include shipping charges and sales tax if appropriate and enclose your check or moneyorder. Individuals must prepay before we can ship your order. Purchase orders are accepted only from bookstores, libraries, universities, and government offices. Please call or write for resale prices.

ORDER FORM

QTY	TITLE	<u>TOTAL</u>
_____	$29.95 Air Conditioning & Refrigeration Technician's EPA Certification Guide	___.___
_____	$14.95 Health Care Job Explosion	___.___
_____	$15.95 The Book of U.S. Government Jobs (5th Edition)	___.___
	* Pennsylvania residents add 7% sales tax	$ ___.___
	* Shipping/handling ($4.25 for first book. $1.00 for each additional book.	$ 4.25
	* Additional Books, ____ x $1.00	$___.___
	TOTAL	$ ___.___

SHIP TO:

NAME: _____

COMPANY: _____

ADDRESS: _____

CITY: _____ **STATE:** _____ **ZIP:** _____

Phone: _____ **Fax #:** _____

❑ I enclose check or money order for $_____
Made payable to: D-AMP PUBLICATIONS

Send orders to:

D-AMP PUBLICATIONS
P. O. Box 1243, Dept. EPA
Moon Township, PA 15108

Phone/Fax 412-262-5578

NOTES